PLANNING AN ECODISTRICT:
Integration of Critical Infrastructure Proposed for the Commune of Léogâne in Haiti

The City College of New York, City University of New York 2012

forward

The not-for-profit Global Energy Model Institute (GEMi), www.globalenergymodel.org, works with developing nations to implement decentralized, low-carbon, cleaner energy systems with proven reliability and resiliency in order to promote self-sufficiency, economic growth, local control and affordability with environmental and cultural protection over the long-term.

The Ecodistrict represents a powerful idea for sustainable development. The CCNY class project explored the application of this concept in Léogâne, Haiti, an area that had been devastated by the earthquake. The integration of sectors necessary for successfully realizing this concept showcases both the strengths of GEM and the Sustainability Masters program. As an interdisciplinary project, the students had to work in groups with multiple specialities to develop projects outside their scope of expertise. The collaborative spirit of GEM is showcased in the work of the students. While GEM is developing the Ecodistrict concept, the representation of Léogâne is an academic exercise only, not intended for implementation.

Daniel E. Lemons
CEO, Global Energy Model Institute

Special thanks to CCNY Professor Michael Piasecki for his support

SUS 7400 A Case Studies in Sustainability Spring 2012
The City College of New York, CUNY

Students: Maria Bueno Rosas, Nuri Celikgil, Steven Cummings, Jie Gu, Julia Ivleva, Priya Kacker, Dania Khan, Danish Kinariwala, Heather Korb, Jessica Mauricio, Ariel Miara, Lisa Morasco, Ayan Owens, Ana Pena, Daniel Plaat, Christopher Sedita, Caleb Stratton, Artemis Velivasaki

Teaching Assistants: Miriam Ward, Caleb Stratton
Professor Hillary Brown FAIA

Hillary Brown, Miriam Ward, editors

Issued September 1, 2012   All Rights Reserved © 2012

table of contents

3	background
4	project information
5	overview: integrated upland infrastructure
6	renewable energy generation and micro-grids
7	wind turbine energy
9	solar photovoltaic energy
11	pumped-storage hydroelectric energy
13	river stabilization
15	habitat and biodiversity restoration at the rouyonne and momance river deltas
17	storm water management, flood mitigation and top soil stabilization
19	riparian buffers: habitat and biodiversity restoration along the rouyonne and momance rivers
21	permaculture ecovillage
23	integrated service infrastructure overview
25	node 1: community hub
27	node 2: town square
29	node 3: nursing school campus
30	node 4: upland rural node
32	node 5: intermodal hub
33	node 6: road improvements and transfer station
36	node 7: eco-industrial park
38	citations and attributions

background

Haiti Overview

This *Case Studies in Sustainability* **seminar project focused on the Haitian city of Léogâne,** which was at the epicenter of the devastating January 2010 earthquake. With 70% of the buildings destroyed and basic infrastructure systems damaged, the innovative approaches envisioned by the students, inspired by the Global Energy Model (GEM), show how critical public services might be restored or established anew. Many of these systems would be collocated to capture potential synergies across the sectors of energy, waste, water, sanitation, transportation, agriculture, flood control and habitat restoration.

The students' interventions were developed along a transect running from the mountainous upland regions of Léogâne (+/- 600m) to the alluvial plain of the settlement and into the sea. The electrification of Léogâne proposed the use of a basic micro-grid. Energy base load energy is supplied by an upland hydro-pumped storage driven by wind and solar farms, a system with designed-in redundancies. This autonomous power system supports new industry and reduces/eliminates dependence on imported fossil fuel. Multiple energy delivery points are located to support local civic functions (community centers, town market and town center) with collocated internet cafes, water services and a waste collection system. These services will be linked by a network of newly paved roads. Collection of organic (field and kitchen) and plastic waste will be incentivized through rebates at small local stations and delivered to a waste processing site at an eco-industrial park attached to an existing sugar mill. Here an industrial sized biodigestor produce biogas for back-up generation and other uses, with organic fertilizer as by-product.

River stabilization with flood control relies on local materials to restructure eroding banks, provide irrigation channels for farming, capture peak flows upland for additional micro-hydro power generation and remediate hazardous flooding conditions from tropical rain events. Riparian buffering will reinvigorate marginal areas while agriculture, agroforestry, aquaculture, irrigation and new rural settlements (arranged according to the **indigenous 'lakou' or multi-family courtyard pattern) all respond to existing topography** and hydrology. These village areas foster biodiversity, habitat, crops, food and biofuel energy in an area of depleted natural resources.

project information

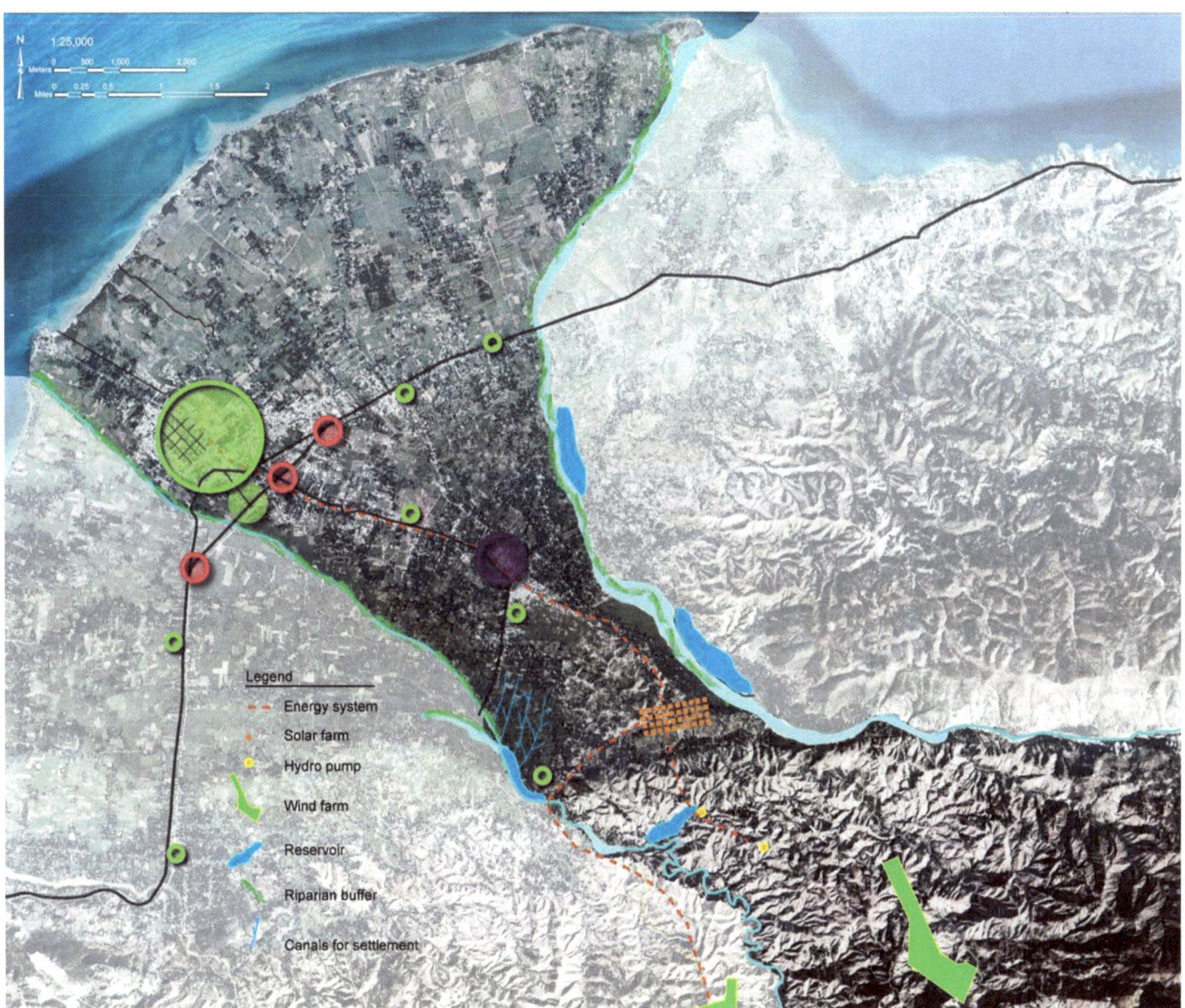

Integration of Critical Infrastructure Overview

This holistic vision for energizing the municipality of Léogâne is based upon decentralized generation of renewable power. Electricity is fed to a series of local micro-grids serving the town center and its outlying settlements. In concert, new public services (water, waste management and transportation) are developed interdependently with energy infrastructure to take advantage of synergies.

The project's unifying objectives are to simultaneously address many of the interconnected problems of Haiti: poverty, malnourishment, pollution, flooding, deforestation, topsoil erosion and loss of biodiversity. Energy, water, sanitation and waste control can promote health, welfare and economic growth across local industry, commerce and agriculture. Through reforestation and riverbank stabilization, the regeneration of degraded ecosystem services will underpin the new energy, water and waste management systems. Infrastructural systems are integrated in such as way as to provide multiple social and economic benefits. Designed-in synergies will help regenerate, rather than deplete environmental resources, support new agriculture, agroforestry, aquaculture and other livelihoods.

overview: integrated upland infrastructure

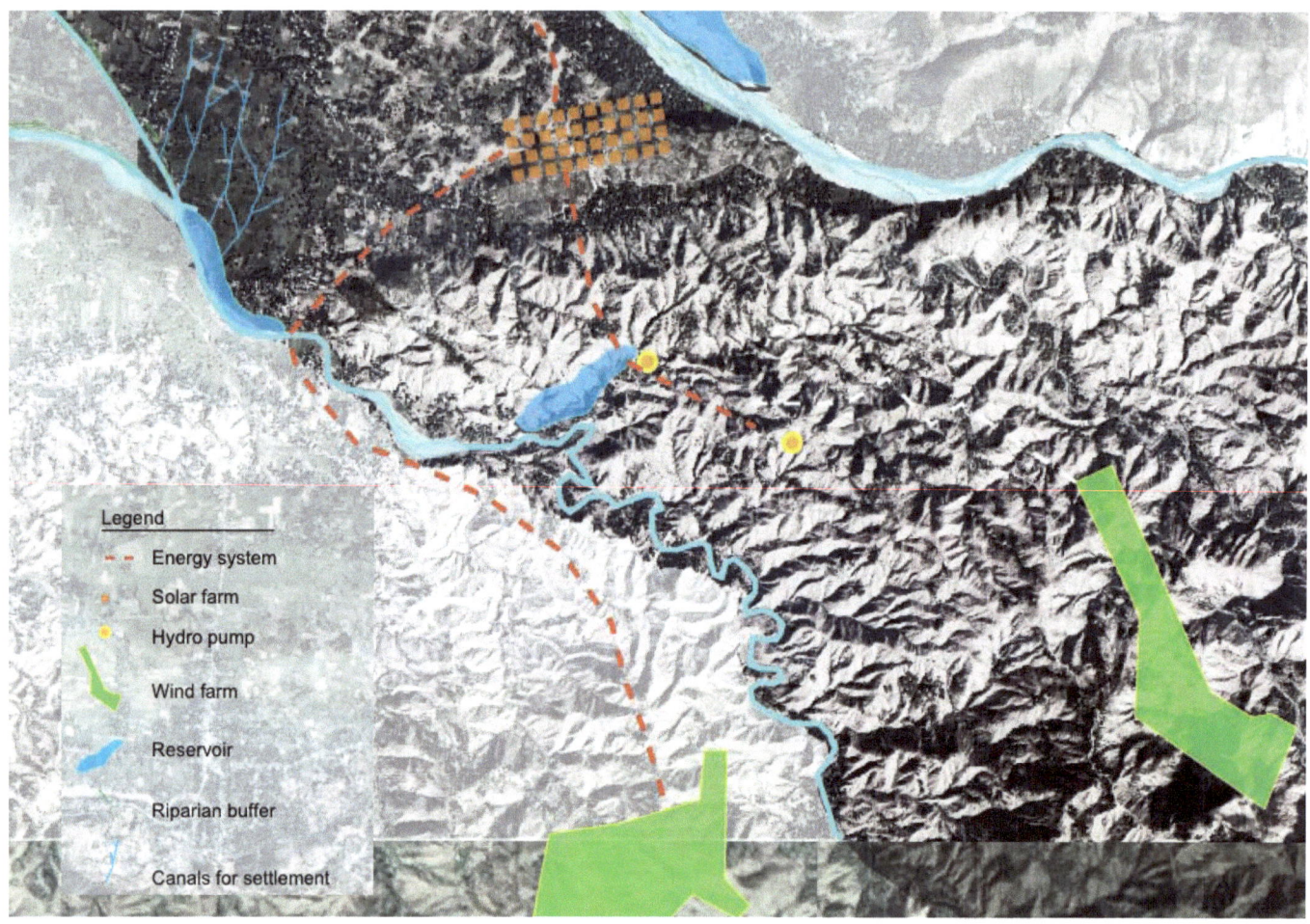

Upland Intervention Sites

problem
To create a local power grid with integrated, sustainable water and waste infrastructures, each infrastructural intervention is designed to foster additional benefits when combined with another sector. The goal is to provide stable utility services for the rebuilding of Léogâne by taking advantage of resources located in a wide corridor running from the mountainous upland to the more populous alluvial plain of the city center.

recommendation
Intermittently available renewable energy (wind and solar) is used to pump water from the lower to the upper reservoir where it is released to generate a base-load for the micro-grid. 10 MW of generation capacity from photovoltaic cells will be installed between the Momance and the Rouyonne Rivers along with 8.6 MW of wind power produced in the upland region. Within this same corridor, installation of riparian buffers, reservoirs, irrigation and delta controls will stabilize and build diversity into a degraded landscape. Combined, these will create self-sustaining upland settlements that enhance quality of life while repairing necessary ecosystem services.

synergies across sectors
Responding to unique aspects of climate, hydrology, topography and land cover, while maximizing output and achieving compound benefits, each sector is designed to foster exchanges (energy, water or resources) with another. This holistic approach is applied to energy generation, environmental stabilization, an eco-industrial park, as well as to agriculture and economic development.

challenges
Adjusting land ownership and creating easements requires neutral, stable governance brokering the complex transactions that allow for placement of infrastructure assets and distribution. An understanding of local values and culture, coupled with local education, are the cornerstones needed for balancing short-term needs for water, food and shelter with long-term stabilization and investment. The logistics of acquiring, installing and maintaining expensive alternative energy infrastructure requires the ongoing presence of trained local technicians.

renewable energy generation and micro-grids

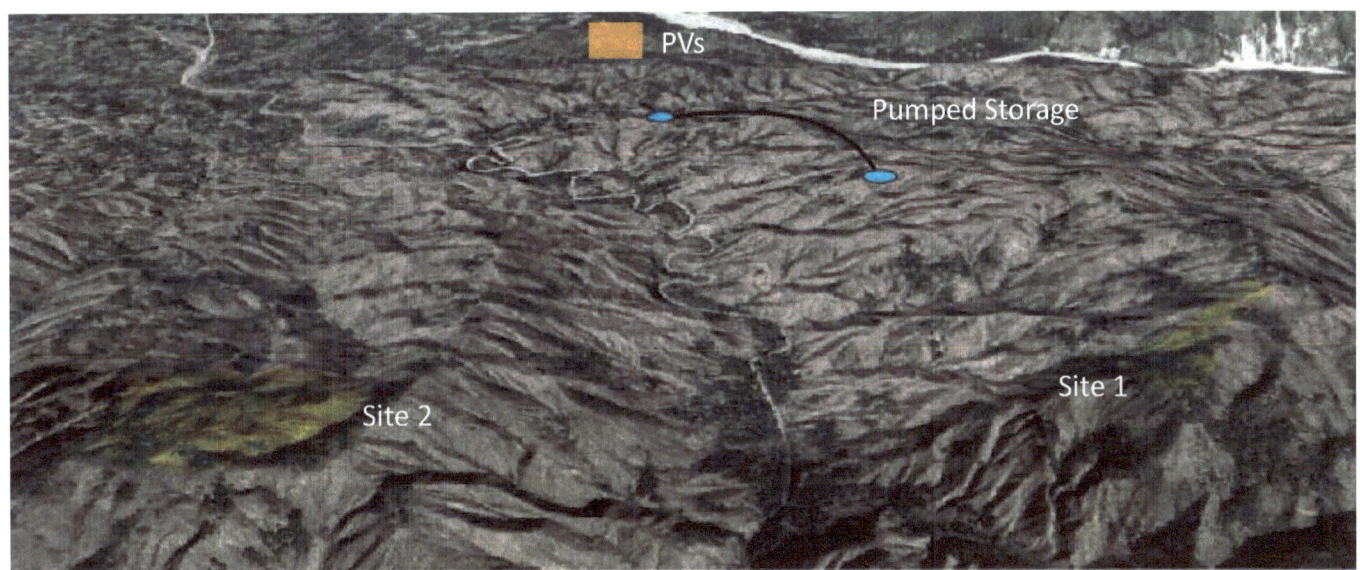

Upland Site Overview

Renewable energy production will provide a controlled source of energy for a given daily time period across a limited grid system. Reliance on renewable resources for domestic, commercial and industrial uses will reduce dependence on imported fuel sources, limit carbon emissions and provide diversification and redundancy. Wind energy production in the mountains coupled with solar energy production in the lowlands, serve a pumped-storage hydroelectric generation station, which balances these dynamic and fluctuating loads.

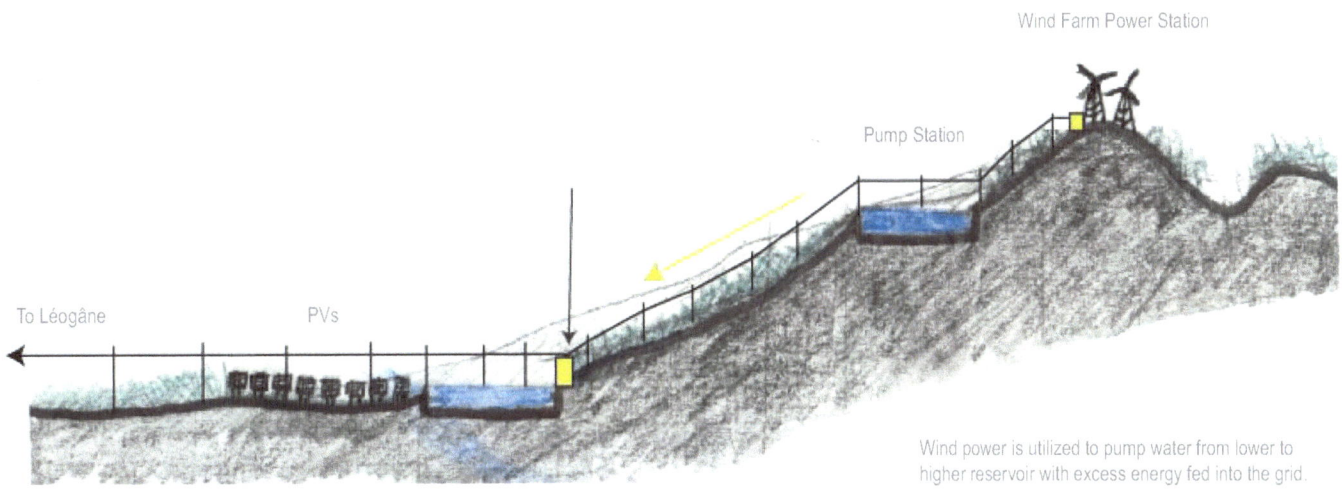

Wind power is utilized to pump water from lower to higher reservoir with excess energy fed into the grid.

Section of Power Systems

wind turbine energy

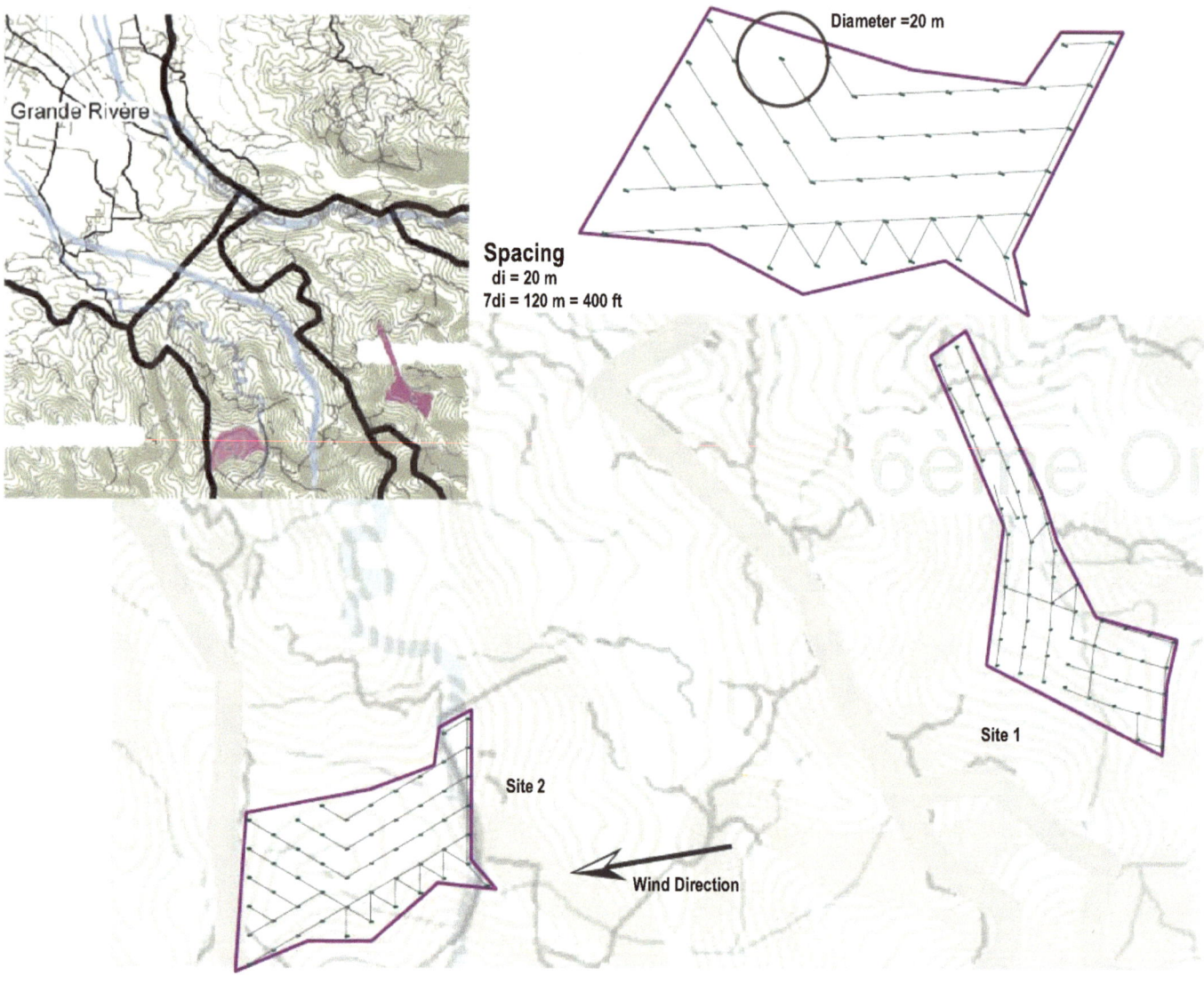

Wind Turbine Site Plan

problem
Sufficient dependable wind energy can be harvested in several locations in the north and west departments of Haiti. In Léogâne however, like much of the rest of the country, this renewable source will not be consistently reliable. The placement of wind farms at higher elevations in the commune (86 turbines with a 8.6 MW capacity) takes advantage of seasonal and diurnal on- and off-shore wind flow. A few small scale technologies, such as the wind-belts, may be useful as backup power to remote sensors and other small stationary devices.

recommendation
Wind energy will primarily be utilized to augment the solar-powered pumped-storage hydroelectric generation system serving Léogâne. Placed on opposite hilltops at elevations of 600 meters, eighty-six 100 kW turbines will complement the main solar farm as primary feed for the pumped-storage system. These medium size turbines, selected in part for easier transport on primitive roads, are supported by lattice towers that provide greater stability in this earthquake-prone area.

synergies across sectors
Wind energy in Haiti can be utilized on a micro-scale and, where sufficient, it can complement solar farm installations during the night. While dedicated to the pumped-storage system, this renewable power system may be required to supplement service on cloudy days.

challenges
It will be necessary to restructure/reallocate land ownership in the upland hills. Some of these areas are highly deforested and eroded so soil stabilization must accompany the relatively costly installation of turbines and access roads.

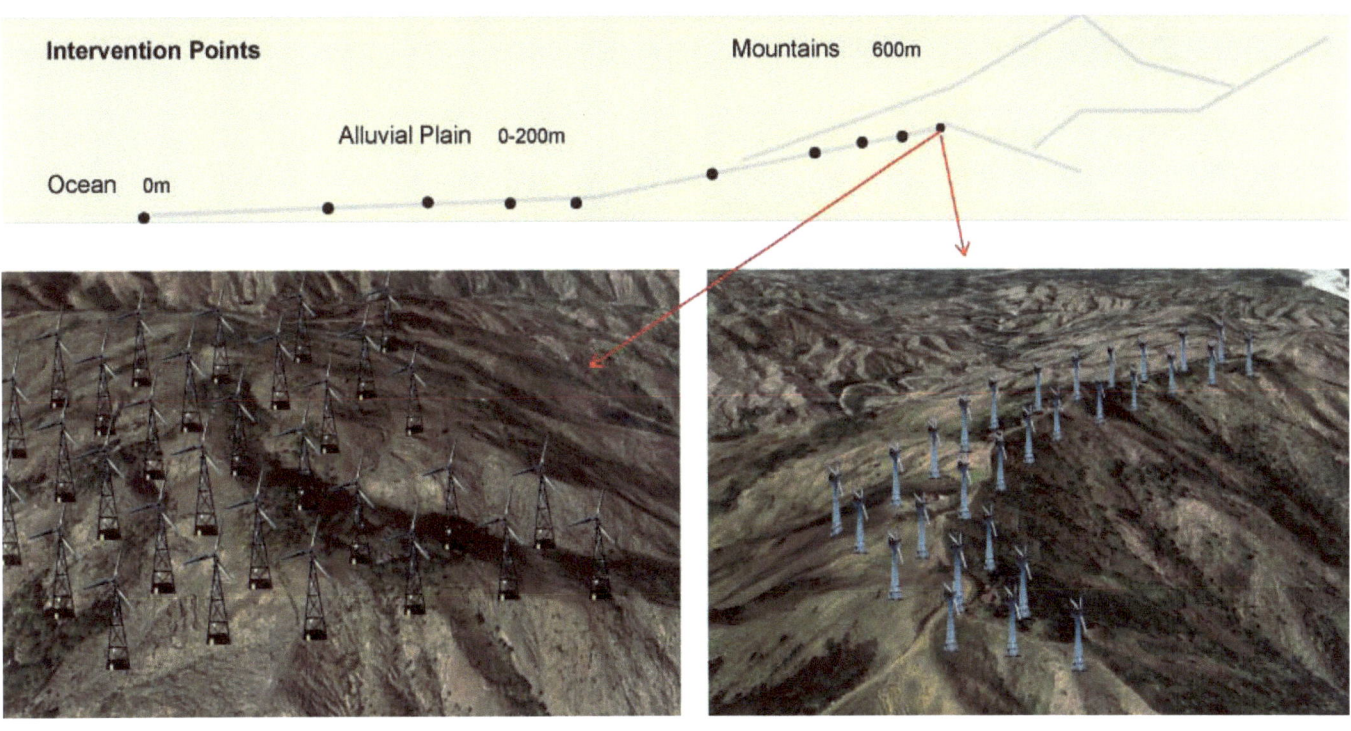

Wind Site 1
41 Turbines = 4.1 MW

Wind Site 2
45 Turbines = 4.5 MW
Both Sites: 86 Turbines = 8.6 MW

Issues
- Risk of Hurricane and earthquake damage
- Switchback road needed to construct wind farms
- Overall low wind speed in the area
- Tests of upland wind patterns needed for most effective placement
- Wind in mountainous areas are hard to predict

Benefits
- Fast installation
- Low maintenance
- Reduced pollutant emissions
- Lattice structure is resisted to against dangerous loads

The Turbine Specifications: 100 kW KCS56

Cost	$50,000 - $300,000 per unit
Generator	Induction 480 VAC 60Hz 3 phase
	Rated output of 100 kW
Transmission	2 stage helical gearbox, 25:1 ratio
Rotor	1800 rpm output, 72 rpm input
Blades	Variable pitch blade design
	Diameter: 60 ft (18 m)
	Swept area: 2463 ft^2 (230 m^2)
Tower	Fiberglass: 27 ft
	Lattice, 80 ft
Control System	Web based automated

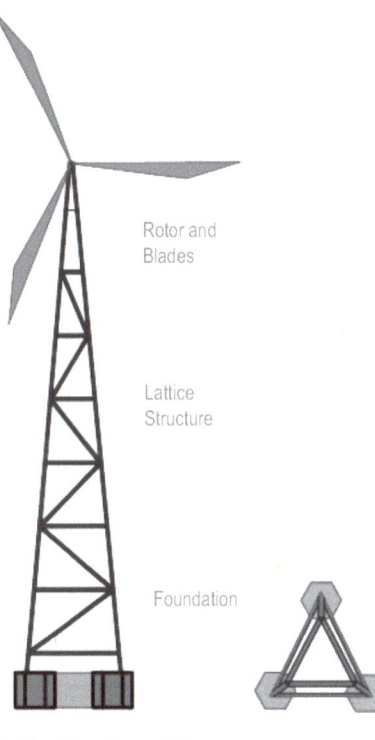

Turbine Elevation and Plan

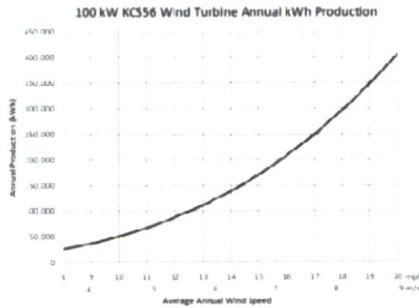

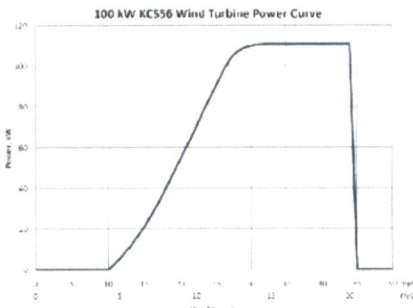

solar photovoltaic energy

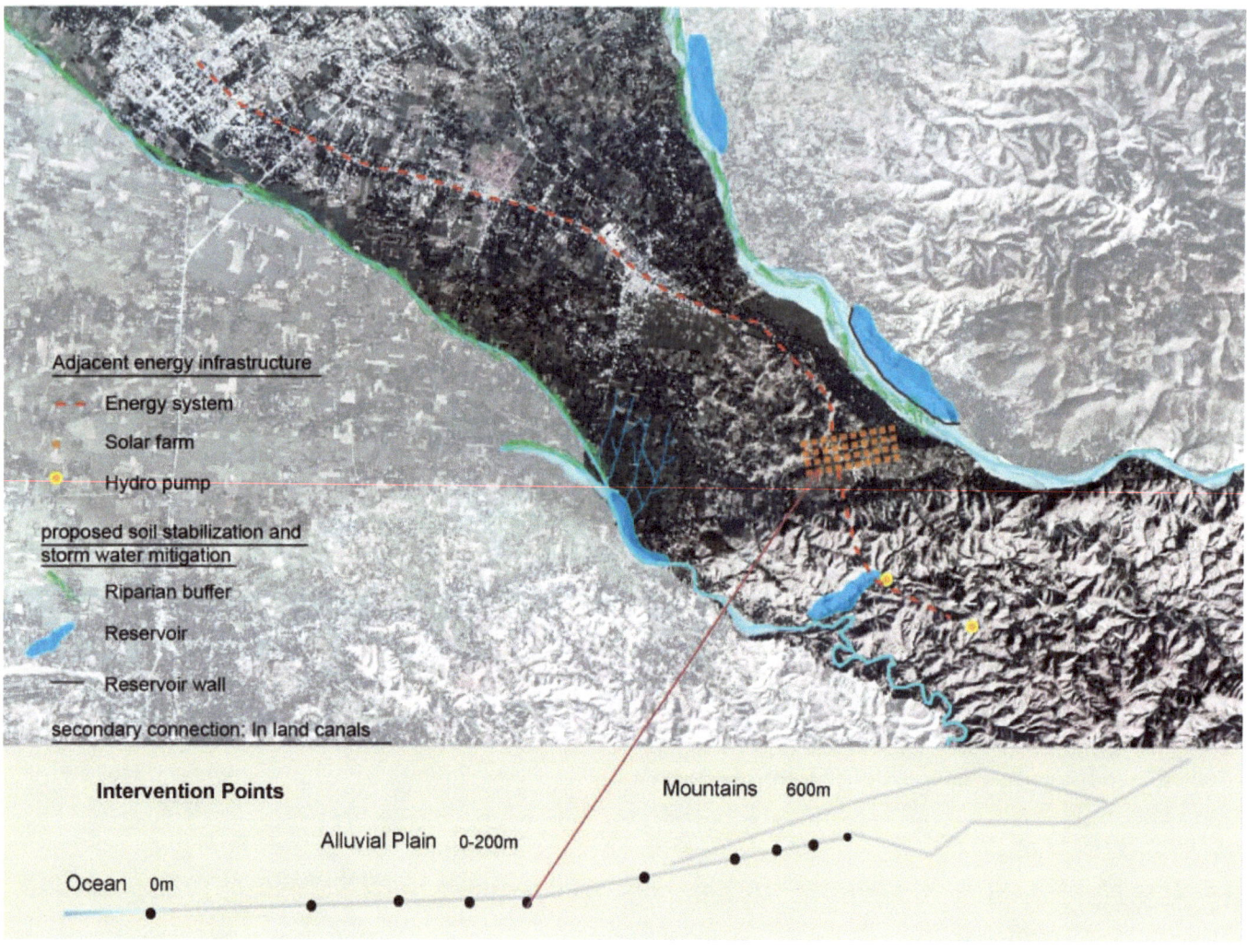

40 Acre Solar Farm

problem
Haiti is located in a prime latitude for photovoltaic energy, yet to date has very little solar capacity and no industrial-scale production of PV panels. Energy generated by solar farms and pumped-hydro storage would reduce fossil fuel emissions while increasing base load capability.

recommendation
A solar farm of 10 MW of generation capacity comprised of 44,000 230 watt commercially available panels will provide electricity during the 11-13 hours of daylight. It will be dedicated to power pumps that drive piped water from a lower reservoir to an upper one. Water stored in the upper reservoir will be released on demand, flowing through turbines to generate a controlled base load. Any excess solar power not needed for water pumping will augment the grid dedicated to Léogâne town center.

synergies across sectors
The solar farms are designed to integrate and not displace agriculture, effectively achieving a kind of 'inter-cropping'. The arrays are located at such a height as to partly shade some of the crops grown under the panels and partially within the service pathways. Rainwater flowing off the panels will be collected in a gutter system for diversion to a cistern and thence to drip irrigation for the collocated crops. The pumped-storage hydroelectric system driven by the complementary sources of solar and wind power will stabilize load production for approximately 8 daytime hours of energy distribution. This reliable service will help spur commercial development and create a Haitian labor market for national goods and services as well as export markets.

challenges
Lack of transportation infrastructure is an impediment to erecting the solar/wind pumped-storage hydroelectric system. The upland corridor identified for the energy generation has little in the ways of roads and no existing electrical transmission cables. A technically trained labor force will be needed for the local operation and maintenance. Gaining clear title to the land with legal contracts for stable operation and maintenance will also pose a challenge. Security concerns must be considered and addressed at all stages.

Floating Solar Array

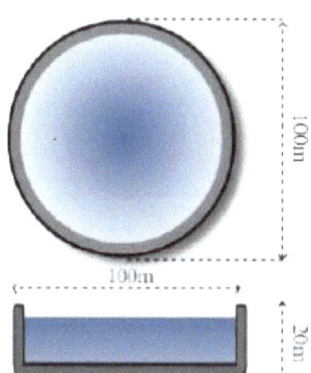

Dimensions of reservoir

Radius:	50m
Area:	7850m²
Height:	20m
Volume:	157000m³

Example of Upper Reservoir

- Place floating PV panels on upper and lower reservoirs
- Reduce evaporation of fresh water, which is hard to replace
- These panels will be more efficient due to cooler temperatures
- Energy produced can also be fed into grid

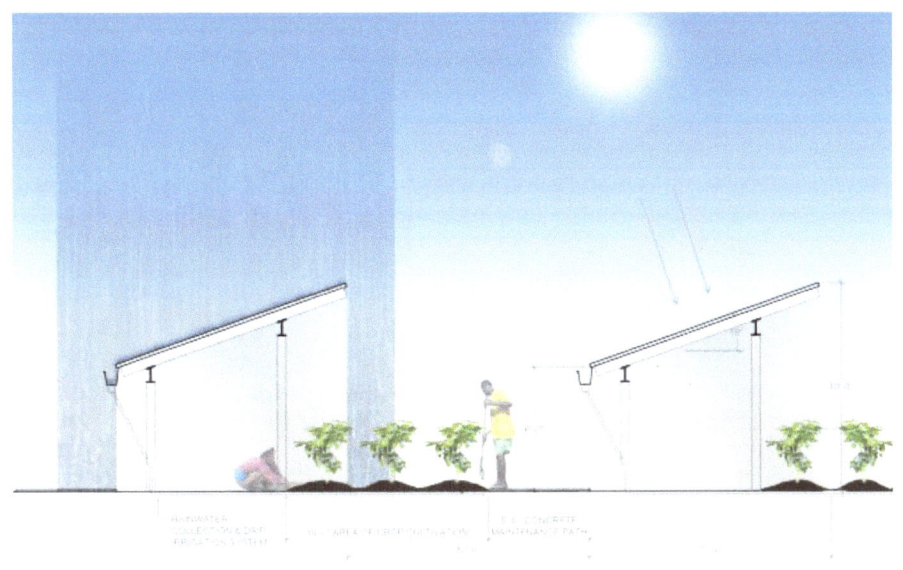

Solar Panels with Agriculture

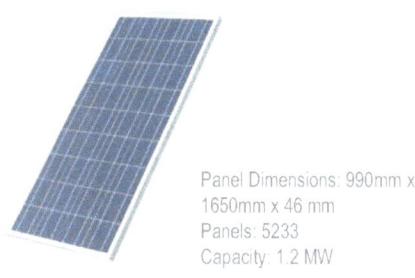

Panel Dimensions: 990mm x 1650mm x 46 mm
Panels: 5233
Capacity: 1.2 MW

- Rainfall and water used to wash panels will be collected in trough.
- Gravity fed to drip irrigation system to supply roots of vegetables grown on-site

pumped-storage hydroelectric energy

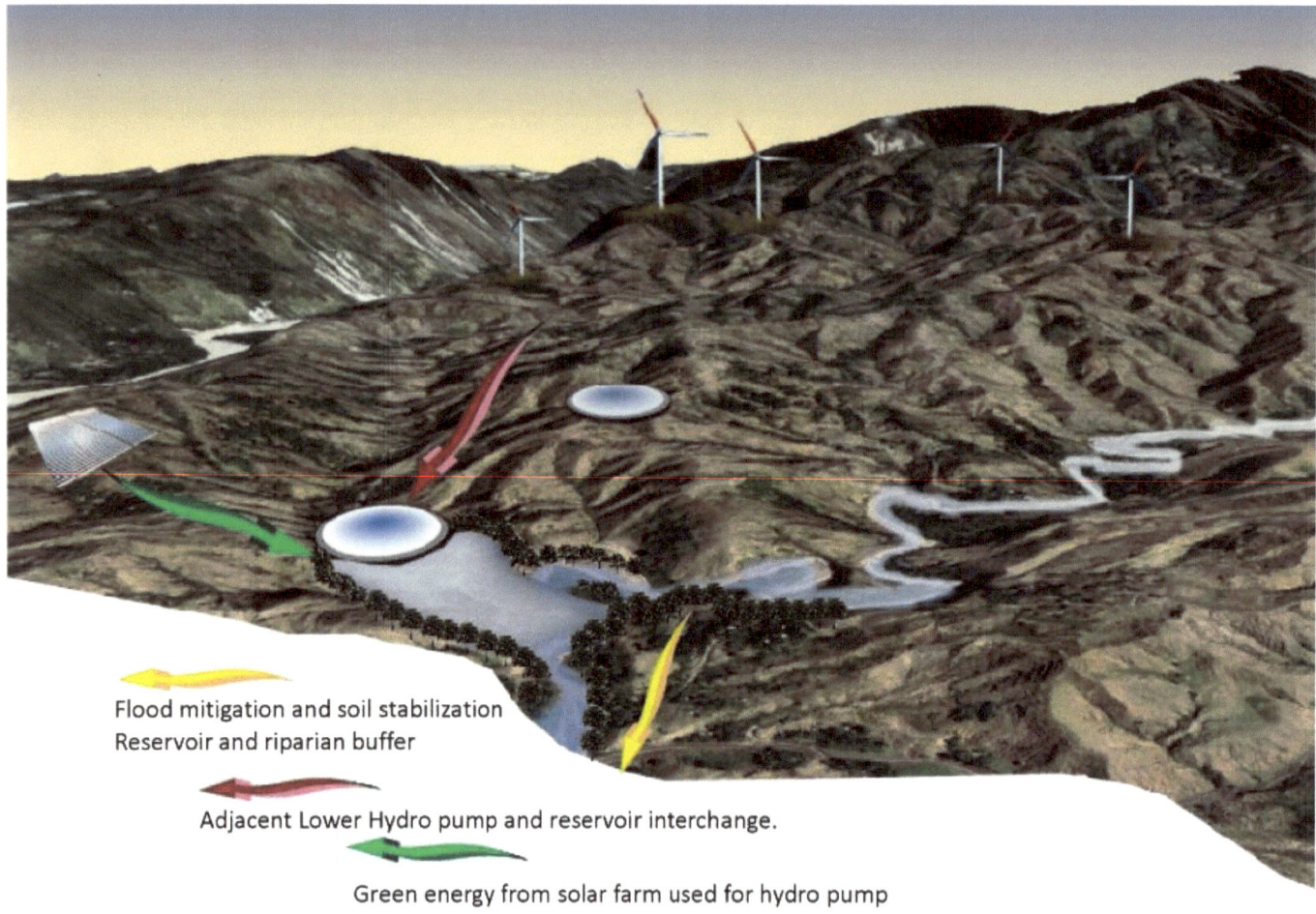

Flood mitigation and soil stabilization
Reservoir and riparian buffer

Adjacent Lower Hydro pump and reservoir interchange.

Green energy from solar farm used for hydro pump

Energy System Co-location Benefits

problem
Most of the existing energy generation infrastructure in Haiti is antiquated, costly to operate and maintain and generates pollutants as well as carbon emissions. While upgrading the capacity of older existing hydroelectric generation is needed to provide an underlying base-load for the future national grid energy, because of negative effects on river systems (sedimentation) and the production of methane from decaying plant matter, these dams are deemed less desirable environmentally. Wind power and solar energy production fluctuates, isn't sufficiently predictable and cannot be stored to provide energy on demand.

recommendation
Pumped-hydro storage (PHS) is a type of power generation generally used for load-balancing. In effect, PHS works like a battery, providing generation capacity for use on demand. Water pumped uphill using wind or solar energy is stored in a higher reservoir and released to flow through turbines on its way to the lower reservoir. The Léogâne system is sized at a capacity of 10 MW, with a 180 meter head. The upper and lower reservoirs each have a water storage volume of 157,000 m^3. Closed-loop flows of water in the PHS system do not interfere with riverine ecology.

synergies across sectors
Stormwater management, flood control, soil stabilization and reforestation are necessary prerequisites for the 'energy corridor' landscape to support all the components of a PHS system. The large reservoirs (upper and lower) will be partially covered by floating PV arrays that constitute part of the solar farm. This collocation means less agricultural land is displaced; it lowers the temperature of the PV arrays, improving their efficiency and their shading of the water reduces the rate of reservoir evaporation.

challenges
Pumped-storage hydroelectric systems require protection from flood, evaporation and seasonal and long-term climate change-induced drought. To operate effectively, local environmental conditions must be stabilized and weather these meteorological stresses. The reservoirs require a measure of protection against soil and sediment accumulation, a serious problem due to deforestation. Soil protection must precede construction of the PHS system. Pipelines must be protected from corrosion and the water being exchanged will require filtration.

Rain input in Léogâne = 0.47 m³/sec (24 hr rainfall, 100 yr event)
- During the raining season, it will take 92 hours to fill up the lower reservoir
- Collect water from rainfall to replace water lost by evaporation
- Use floating PVs to shield water from evaporation and to generate energy

Pumped Hydro Storage Plan and Elevation

Rendering of Pump Water System

Pumped Hydro System
Capacity: 10 MW
Head (h): 180m
Turbine efficiency: 80%
Turbine discharge:

$$Q = \frac{MW}{9.8 \times 10^{-3} \times h \times eff} = 7.08 \, m^3/s$$

Generation time: 6 hours
Pumping energy needed: 13MW
Overall cycle efficiency: 75%
Upper and Lower reservoirs volume: 57,000 m³ each

Bird's Eye View of Pump Water System

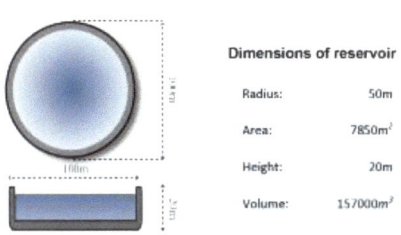

Dimensions of reservoir
Radius: 50m
Area: 7850m²
Height: 20m
Volume: 157000m³

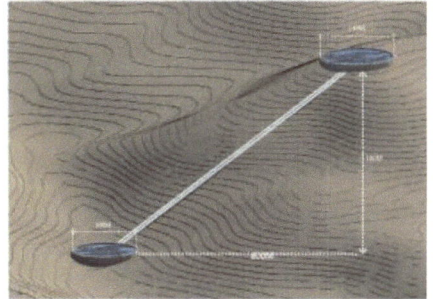

Horizontal and Vertical Distance Between Two Reservoirs

river stabilization

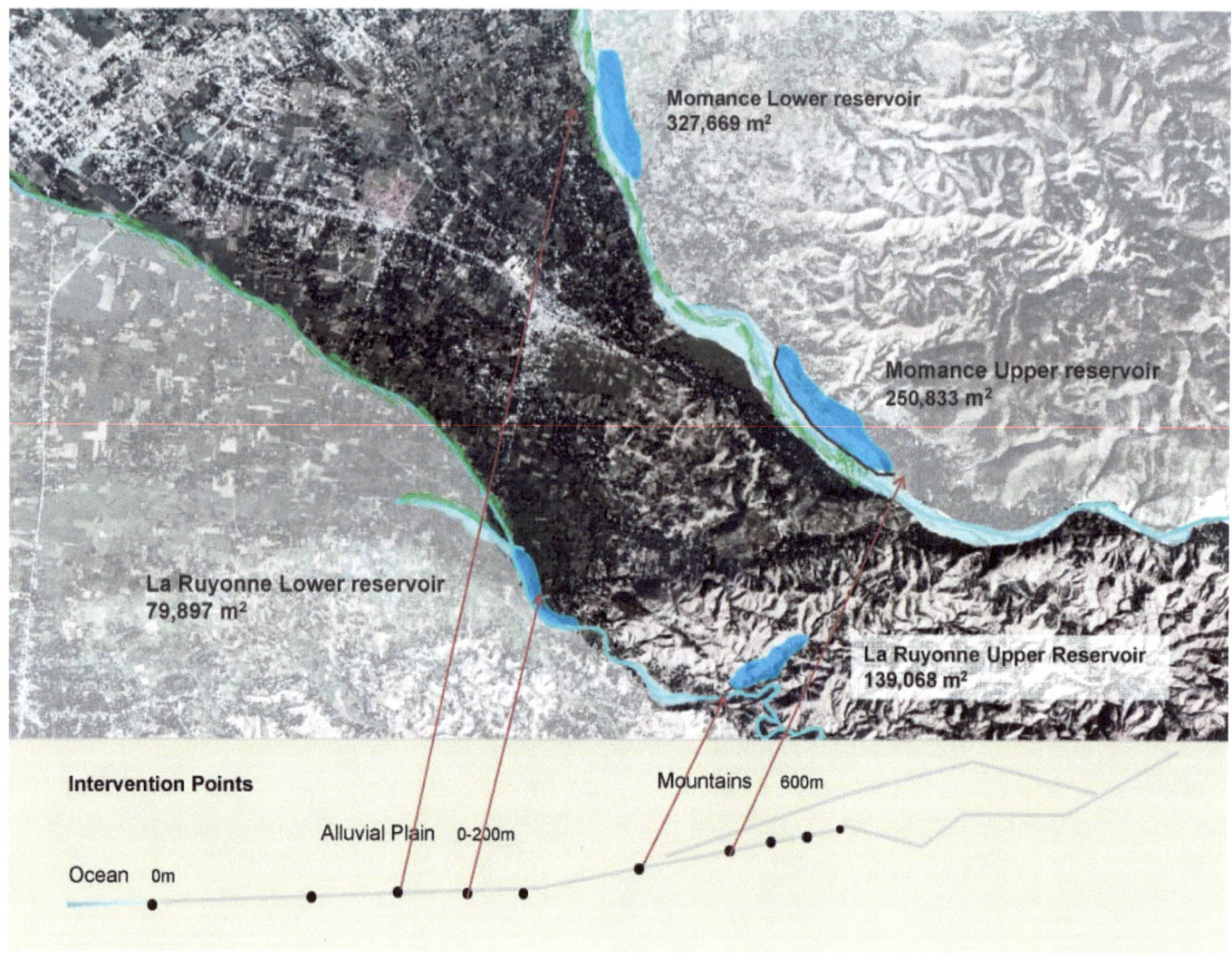

Proposed Reservoirs

River stabilization is focused on the Rouyonne and Momance Rivers, due to the magnitude and frequency of flooding events. Efforts will rely on natural and readily available materials to restore eroding riverbanks, provide irrigation channels for farming, limit upland erosion, capture peak flows upland for power generation and address hazardous flooding conditions during tropical storm events. Upland reservoirs are built to reduce downstream flooding in both rivers; they will supply agricultural irrigation during dry seasons. With small micro-turbines, the dams can be used to power irrigation pumps.

Intervention Points

Mountains 600m
Alluvial Plain 0-200m
Ocean 0m

The intervention points providing flood control, erosion controls, irrigation and energy production.

Momance River Delta

La Ruyonne River Lower Reservoir 79,897 m²
Rain input= 0.27 m³/sec *(24hr rainfall-100yr event)*
Assuming H=10m → **Storage=798,970 m³**
Available water for agriculture in dry seasons

Momance Lower reservoir 327,669 m²
Rain input= 1.1 m³/sec *24hr rainfall-100yr event*
Assuming H=10m → **Storage=3.28x10⁶ m³**
Available water for agriculture in dry seasons

La Ruyonne River Upper Reservoir 139,068 m²
Rain input= 0.47 m³/sec *(24hr rainfall, 100-yr event)*
River input= 3.20 m³/sec *(½ Momance's flow rate)*
Total Input: 3.67 m³/sec
Assuming H=20m → **Storage=2.6x10⁶ m³**
8 days of constant rainfall to fill

Momance Upper reservoir 250,833 m²
Rain input= 0.85 m³/sec *24hr rainfall, 100-yr event*
River input= 6.4 m³/sec
Total Input: 7.25 m³/sec
Assuming H=20m → **Storage=5x10⁶ m³**
8 days of constant rainfall to fill

Reservoir Capacity

habitat and biodiversity restoration at the rouyonne and momance river deltas

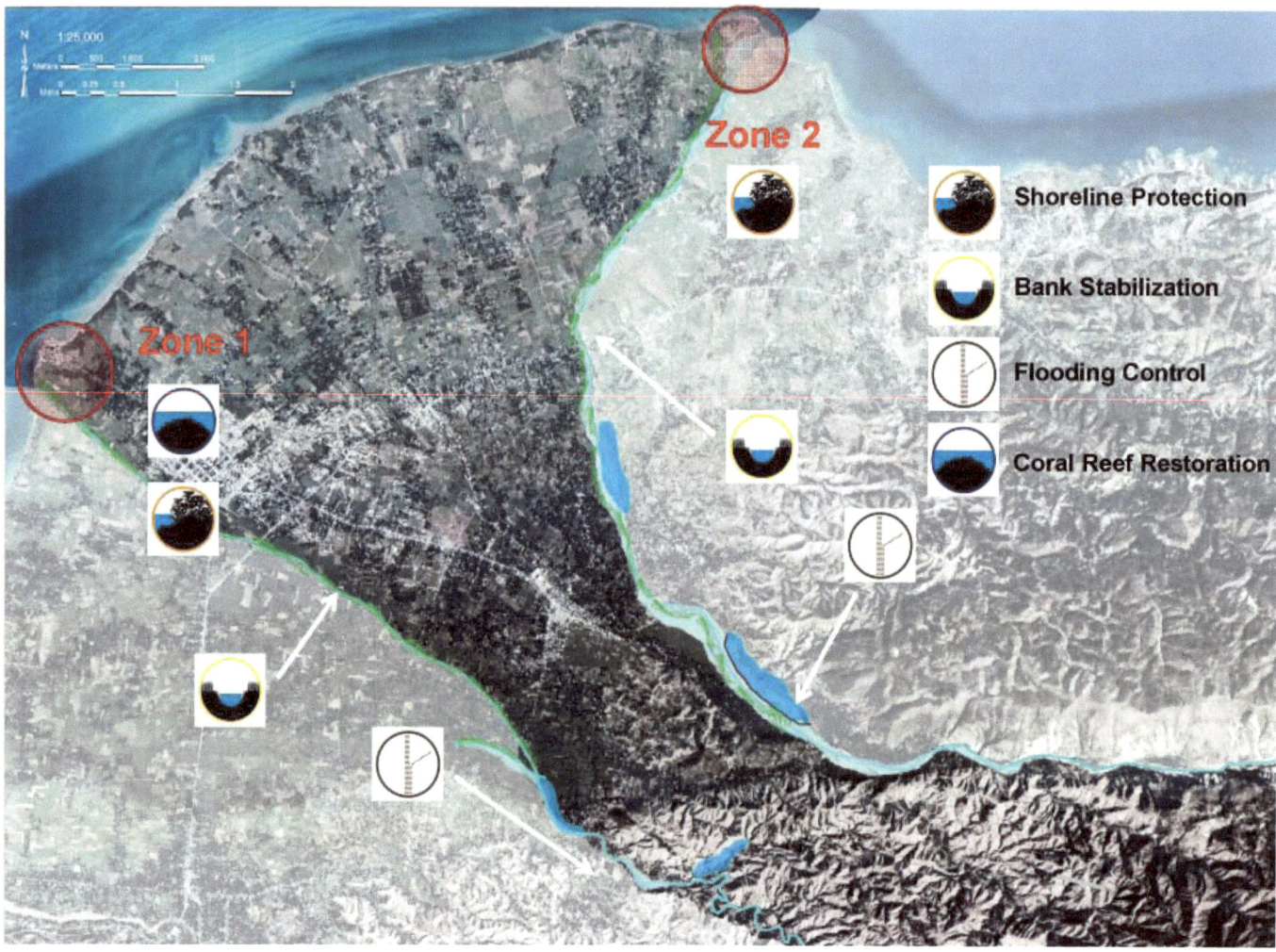

Shoreline Interventions

problem
Significant damage has been sustained in both the Momance and Rouyonne river deltas along some 55 km of Léogâne. While natural causes for habitat degradation include seismic uplift that drains tidal marsh, exposing sub-tidal sea grass and causing loss of coral reefs, anthropogenic causes include massive influx of river sedimentation from the deforesting watershed mountains and harvesting of mangroves for firewood.

recommendations
Restoration of the imperiled coral reefs and mangroves, along with flood mitigation and bank restoration are imperative. For reef rebuilding, one promising artificial platform system has been developed called "Reef Ball". It consists of a perforated concrete form into which plugs of healthy new coral fragments are introduced before being placed on the ocean floor. Mangroves should be planted in biodegradable baskets using a range of salt-tolerant and saline resistant species based on local conditions.

synergies across sectors
Mangrove restoration will help protect the fragile corals implants since they act as filtering system, trap debris and silt, produce nutrients, improve fisheries and protect coastal zones from storms and major weather events. Reforestation and river bank reconstruction will reduce outflows of sediments.

challenges
Management and enforcement of the National System of Protected Areas for Coastal Areas is essential to reduce further mangrove harvesting and overall for investment to succeed. Additionally, both mangrove and coral reef restoration will require trained personnel.

Zone 1: Arial View

Consequences of the Earthquake January 12, 2010: Along 55 km of coastline from Gressier (East of Léogâne) to Port Royal (West Haiti) the seismic uplift 0.64 ± 0.11 m of elevated coral reefs

"Reef Ball" NGO System
Proposed for restoring biodiversity

Reef Ball Mold System

Zone 2: Momance River Delta

Spread river mouth due to deposited sediment because of lack of soil stabilization

storm water management, flood mitigation and topsoil stabilization

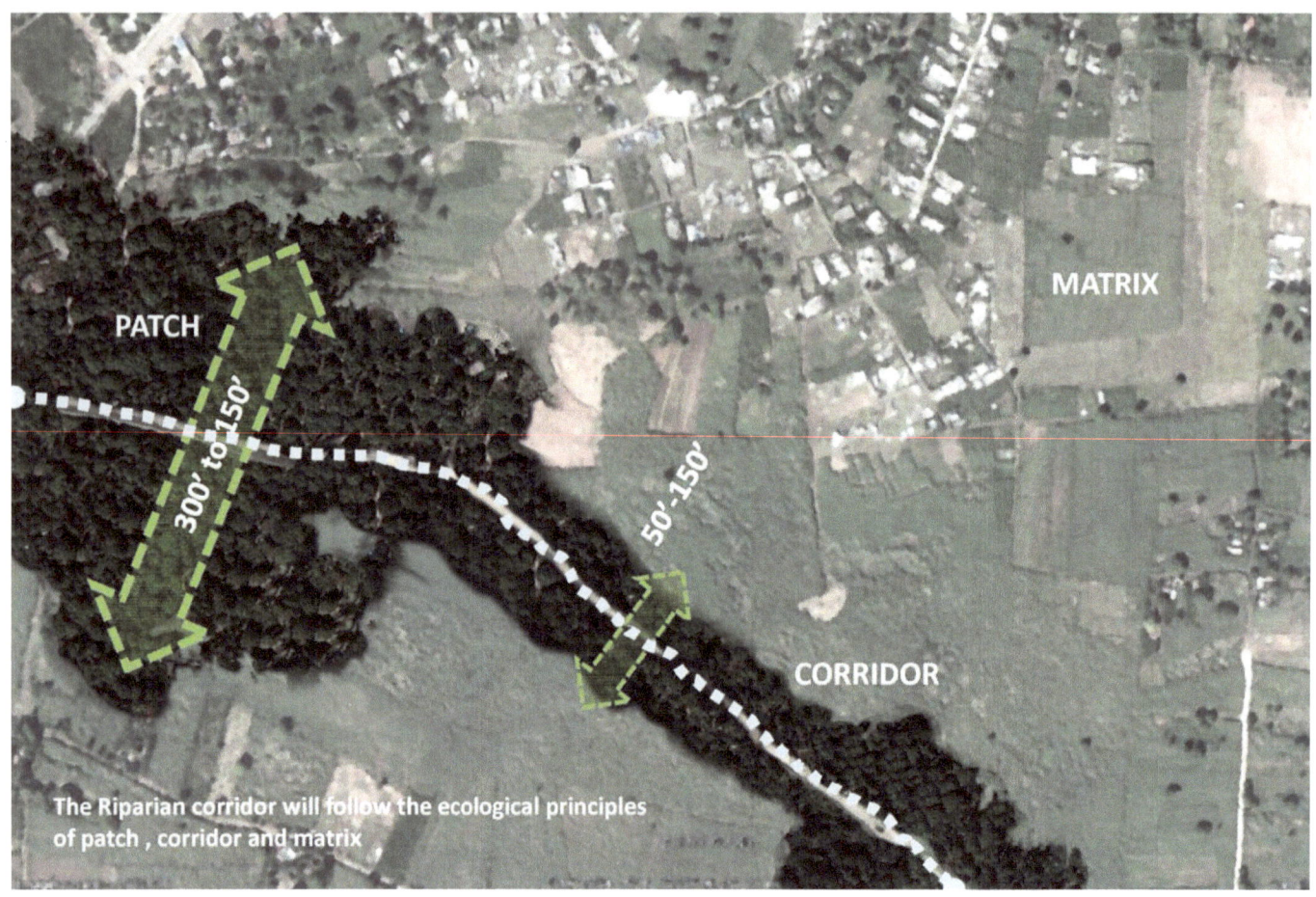

Riparian Strategy: Patch, Corridor, Matrix

problem
Deforestation is a critical issue in Léogâne, as it is in many areas of Haiti. It is severe in mountainous regions, as trees are removed for agriculture and to create lumber or charcoal. Without roots to restrain the soil and absorb stormwater, rivers and streams are overwhelmed during heavy rains. Topsoil loss is endemic, with little vegetation to prevent sediment loss from countryside slopes. This compounds flooding as drainage paths fill with upland soil and debris. Low-lying Léogâne and its estuaries are especially vulnerable to these flood hazards.

recommendation
The placement of 100-foot riparian buffers with excavation for four distinct new reservoirs will mitigate the effects of stormwater runoff. Buffers consisting of trees (over 380,000 estimated), shrubs, vertiver plantings for cash crops and riprap from concrete debris provide flood control and soil stabilization, improving living conditions and increasing yield from agriculture. Creation of new reservoirs to which floodwaters can be diverted (875 m^3 capacity for the Momance and 220 m^3 for the Rouyonne) will allow for the stabilization of both riverbanks.

synergies across sectors
The riparian buffer fosters concentrated vegetation within riparian zones that constitute habitat corridors for diversifying species migration across large areas. Flood management improves agriculture by allowing for replenishment of topsoil and root strengthening. Creation of new upland reservoirs provides flood control, irrigation water and power generation through micro-turbines.

challenges
Barriers to these interventions in Léogâne include trained manpower and local support for labor-intensive upkeep of the riparian buffer plantings and reservoir and irrigation maintenance release gate during stormy and dry seasons. Easements or land acquisition will be necessary.

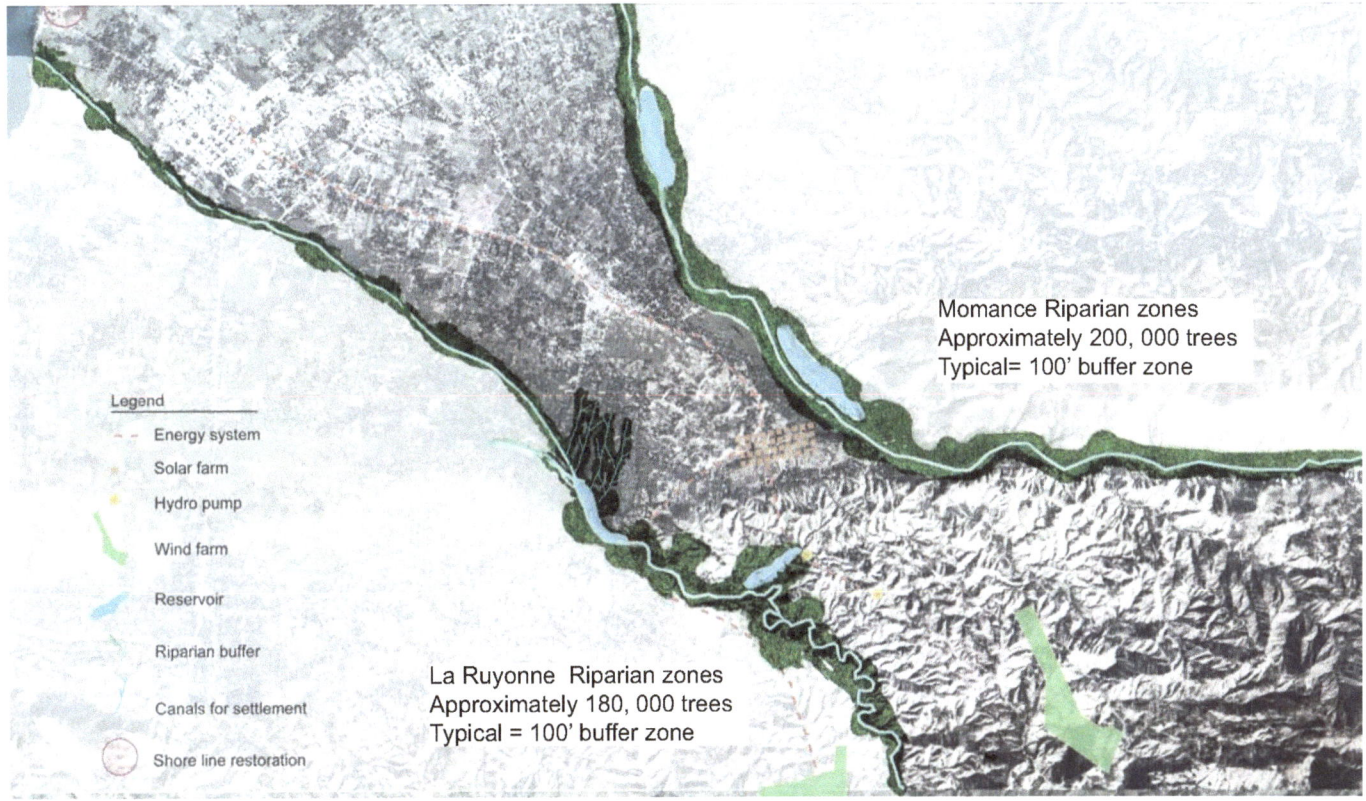

Riparian Buffer Layout and Illustration

Ecological Services

Flooding and Temperature Control

- Diversity and concentration of vegetation within riparian zones provides habitat for a variety of animals and insects
- Long, continuous strips of riparian buffer can also provide habitat corridors for the migration of species across large areas
- Absorb flood waters that would otherwise damage adjacent urban, forest or agriculture lands and contribute to stream bank erosion
- Filtrates phosphorus and other pollutants from runoff
- Shade form plants and trees within riparian zones helps regulate water temperature
- Maintain biological diversity

riparian buffers: habitat and biodiversity restoration along the rouyonne and momance rivers

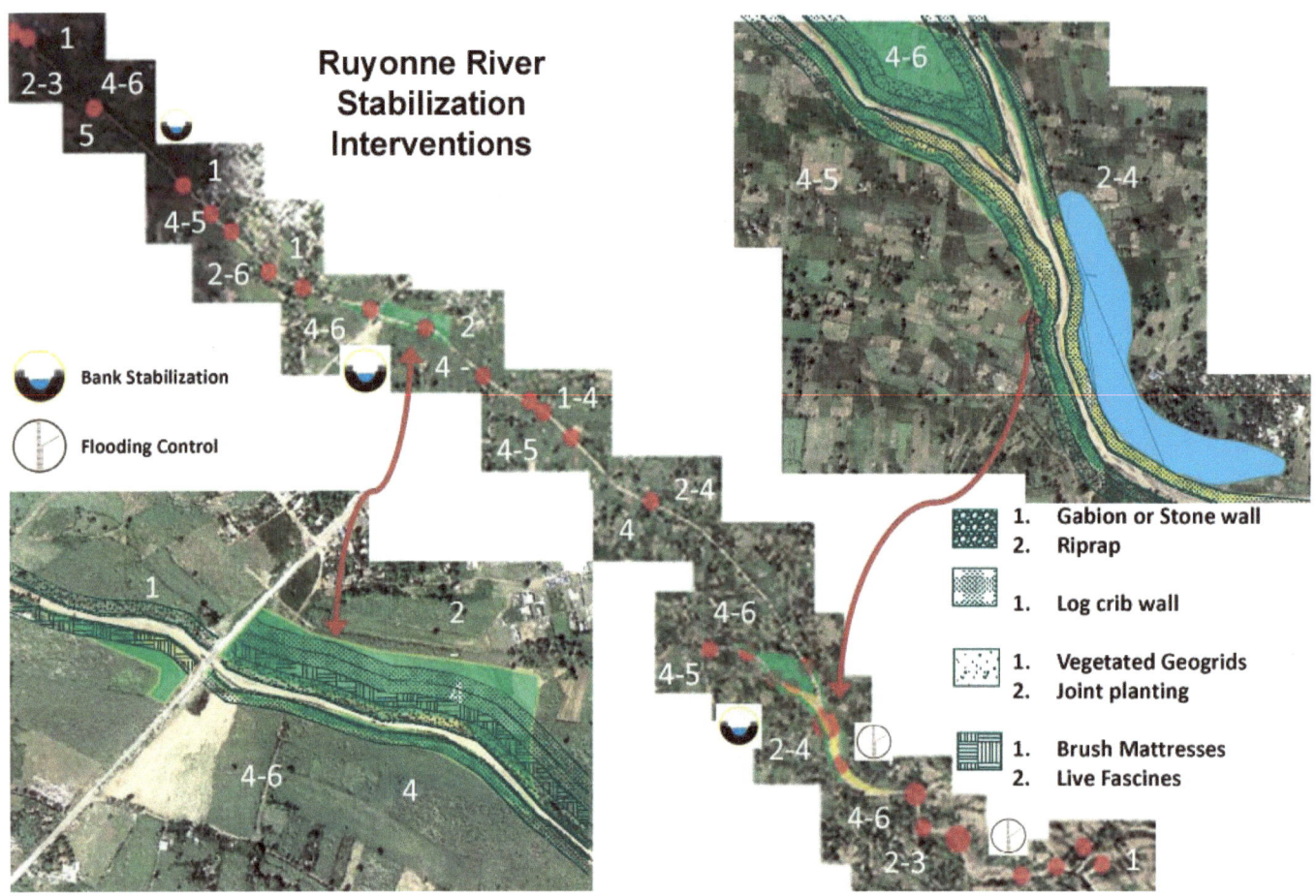

Application of Bank Stabilization and Flood Control

problem
In Léogâne, riparian areas with native vegetation were formerly oases of ecological diversity. Human land-use practices degraded these corridors, opening the way for riverbank erosion and flooding, loss of agricultural lands, damage to human settlements and rapid sedimentation in the river deltas. These valuable riparian zones must be the urgent focus of ecological restoration efforts.

recommendations
A restructuring of land holdings along these rivers, including 'no-trespassing' conservation zones may be required to foster restoration and long-term management of these corridors. The creation of vegetated and forested buffer strips and patches (between 50 to 300 feet across) along river/stream banks armors the bank structure, absorbs flood waters and slows erosion, improves water quality (pollutant removal, temperature moderation) downstream and creates habitat corridors.

synergies across sectors
In addition to the structural and ecological services rendered by the restored riparian buffers, trees and plantings integrated using 'agro-forestry' models (intercropping of trees and food plants) can produce food, fuel and create jobs. Fruit production, timber from Jatropha, vertiver and coir industries can be integrated with adjacent irrigation channels needed to boost agricultural productivity.

challenges
Restructuring and reallocating land ownership along the river corridors will require political and administrative support with either financial or social incentives. There is an overarching need to change environmentally unsound agricultural practices and inculcate an appreciation of a 'shared' habitat. Trade-offs need to be made between slow-growing native species vs. fast-growing exotics when stabilizing the riverbank.

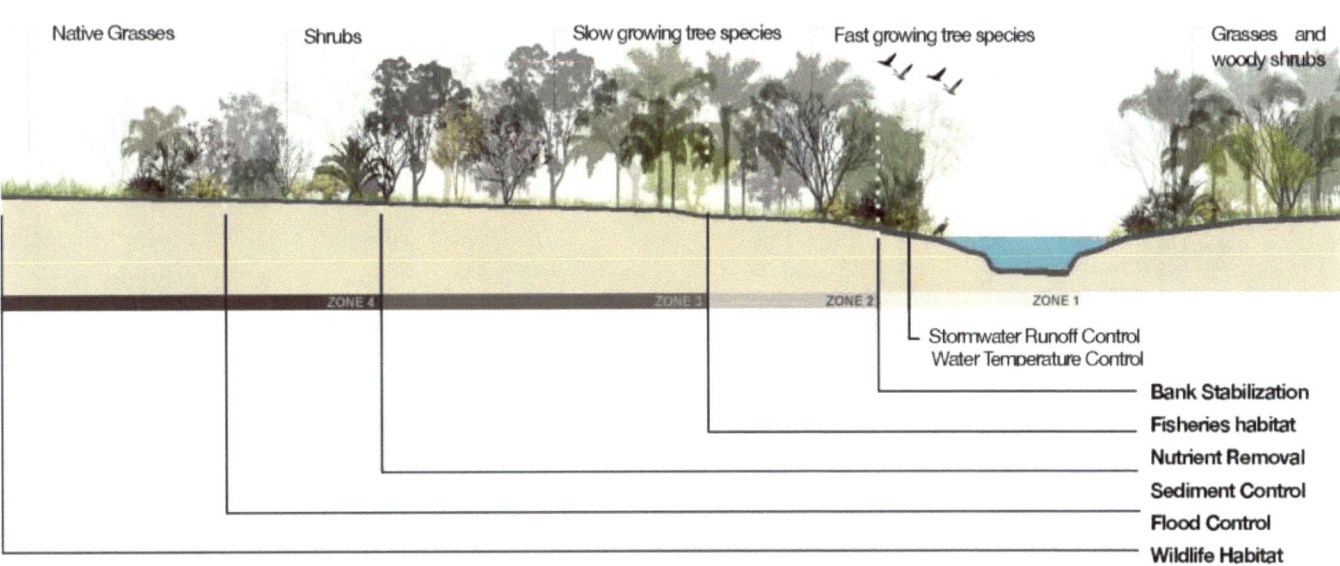

Riparian Buffer Physiology

Haley Heard, MIT

ZONE 4 150m	ZONE 3 100m	ZONE 2 50m	ZONE 1 0m
Zone 4 functions to intercept and dissipate the energy of surface runoff, trap sediment and agricultural chemicals in the surface runoff, and provide a source of organic matter for soil microbes that can metabolize nonpoint source pollutants. Native grasses with a uniform cover that has dense, stiff stems provides a highly frictional surface to intercept surface runoff and facilitate infiltration. (Dabney et al. 1993)	Zone 3 consists of a strip of tall grasses or herbaceous cover to spread and filter runoff which may be transporting sediment, nutrients and pesticides off urban land, cropland, or erosive or sparsely vegetated areas. The establishment of this zone is critical where the control of sediment, nutrient, pesticide or non-point source pollution is necessary, as is the case in urban and agricultural situations.	Zone 2 contains trees and shrubs and other vegetation needed to filter runoff and absorb nutrients and pollutants. Dominant vegetation consists of existing or planted trees and shrubs suited to the site and purpose. Forest management and tree harvesting is permitted as long as the purpose of the zone is not compromised. Tree harvesting allows the landowner to maintain the land's productive value while providing water quality benefits at the same time.	Zone 1 is adjacent to the water and contains trees and shrubs needed to provide shade, insect habitat, bank stability and large woody debris for in-stream habitat. The complex root structures of woody plants are highly desirable for holding soil in place, improve bank stability and to maintain a natural riparian ecology.

ZONE 4
- Jatropha curcas
- Agave sisalana
- Pineapple
- Sugarcane

FRUIT TREES
- Avocado-*Persea americana*
- Mango-*Mangifera indica*
- Coconut-*Cocos nucifera*
- Citrus spp. (lemon/lime/orange)
- Plantain
- Papaya
- Cacao

ZONE 3
- Mahogany - *Swetenia mahagony*
- Spanish Cedar-*Cedrela odorata*
- Logwood –*Haematoxylon campechianum*
- Bayawonn-*Prosopis juliflora*
- Bwa Soumi-*Cordia alliodora*
- *Coccothrinax argentea*
- *Sabal causiarum*
- Chenn –*Catalpa longissima*

ZONE 2
- Mesquite – *Prosopis juliflora**
- Moringa*
- Leucaena –*Leucaena leucocephala**
- Neem – *Azadirachta indica**
- Cassia-*Senna siamea**
- Simarouba sp.
- Royal Palm -*Roystonea borinquena*
- Mango – *Magnifera Indica*
- Bwa Ple- Colubrina arborescens

ZONE 1
- Vetiver grass-*Vetiveria zizanioides*
- Napier grass- *Pennisetum purpureum*
- Guinea Grass -*Panicum maximum*
- Lemon Grass (*Cymbopogon citratus*)

permaculture ecovillage

Proposed Permaculture Ecovillage Compound for Rural Upland Site

problem
Permaculture is a design and ethical system for ecological living, integrating food, shelter, energy, climate, water and soil. The problems of Haiti—poverty, malnourishment, pollution, deforestation, erosion, degradation of coastline, riverbanks and lost biodiversity—are interconnected and beg permaculture solutions. These are especially effective in stabilizing and reinvigorating marginal areas.

recommendation
The 'ecovillage' is a proposal to foster sustainable livelihood, environmental stewardship, social bonding, economic stability and balanced nutrition. In the ecovillage, finger ponds are excavated adjacent to the pond reservoirs and canals during the dry season. They fill during the flood cycle, thereby trapping fish as the flooding recedes. During the following dry season, the ponds are enriched with manure while fish are cropped. Adjacent vegetable gardens are watered from ponds and fertilized with pond sediments following the fish harvest.

synergies across sectors
Fish farming and rice/fish co-farming are integrated components of this vision, assisting with poverty alleviation, increased protein consumption, protection of the environment and natural resources and livelihood provision. A waste-to-food cycle that includes exchanging nutrient waste from/to agriculture and aquaculture processes eliminates waste and pollution.

challenges
Education is necessary to empower fishermen, farmers and households to protect their own natural resources. Ecovillage implementation must respect property rights and work with area inhabitants to create sustainable living situations. Detailed knowledge of sustainable water management is imperative for a thriving fish industry. Local culture is dominant and will frequently work against successful resource management. Current practice in waste disposal, latrines, and charcoal cooking are three areas likely to present major challenges.

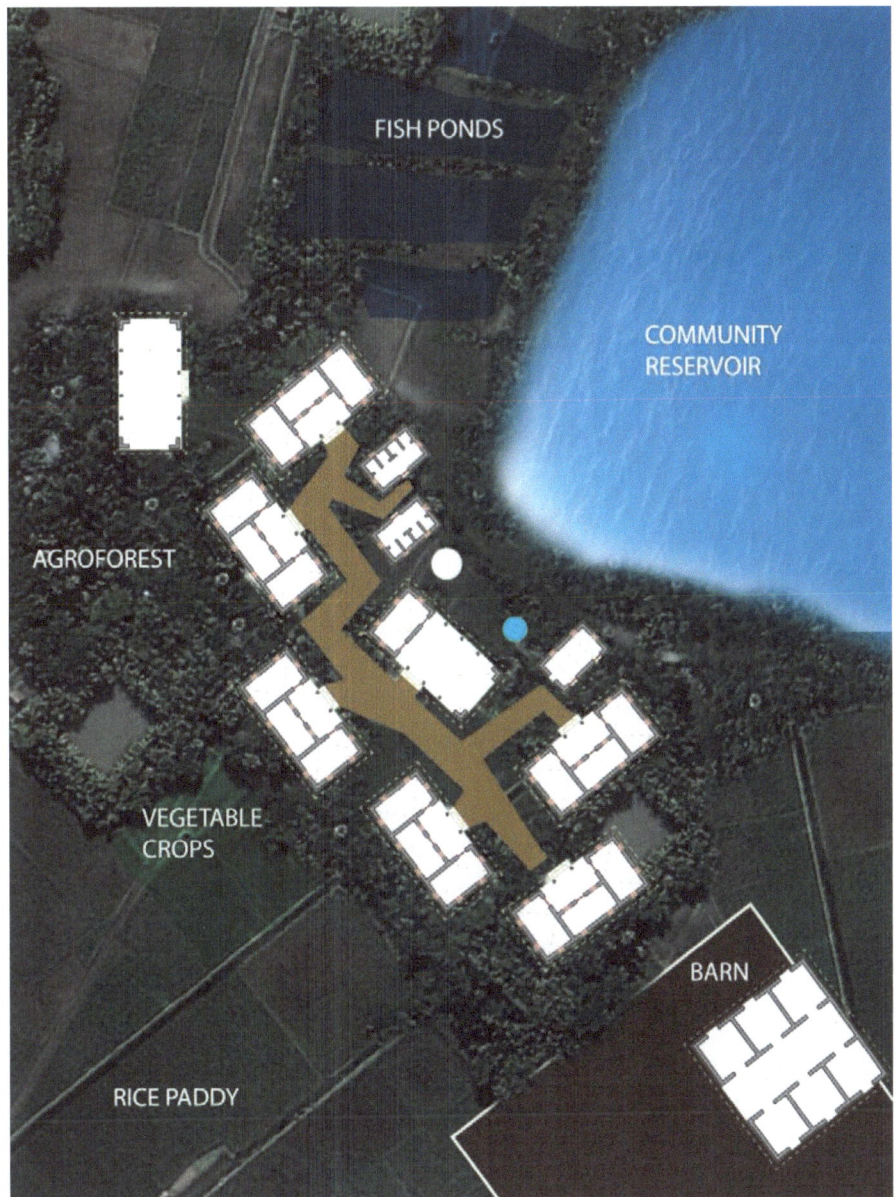

Permaculture Ecovillage Compound

- Typical landholding per family = 1 ha
- 130 ha productive land = 130 families
- Typical family size = 7 persons
- Community size ≈ 900 residents

Permaculture Site Overview

Compounds Arrayed Around Man-Made Water Bodies

Compound Layout

Compound Perspective

integrated service infrastructure overview

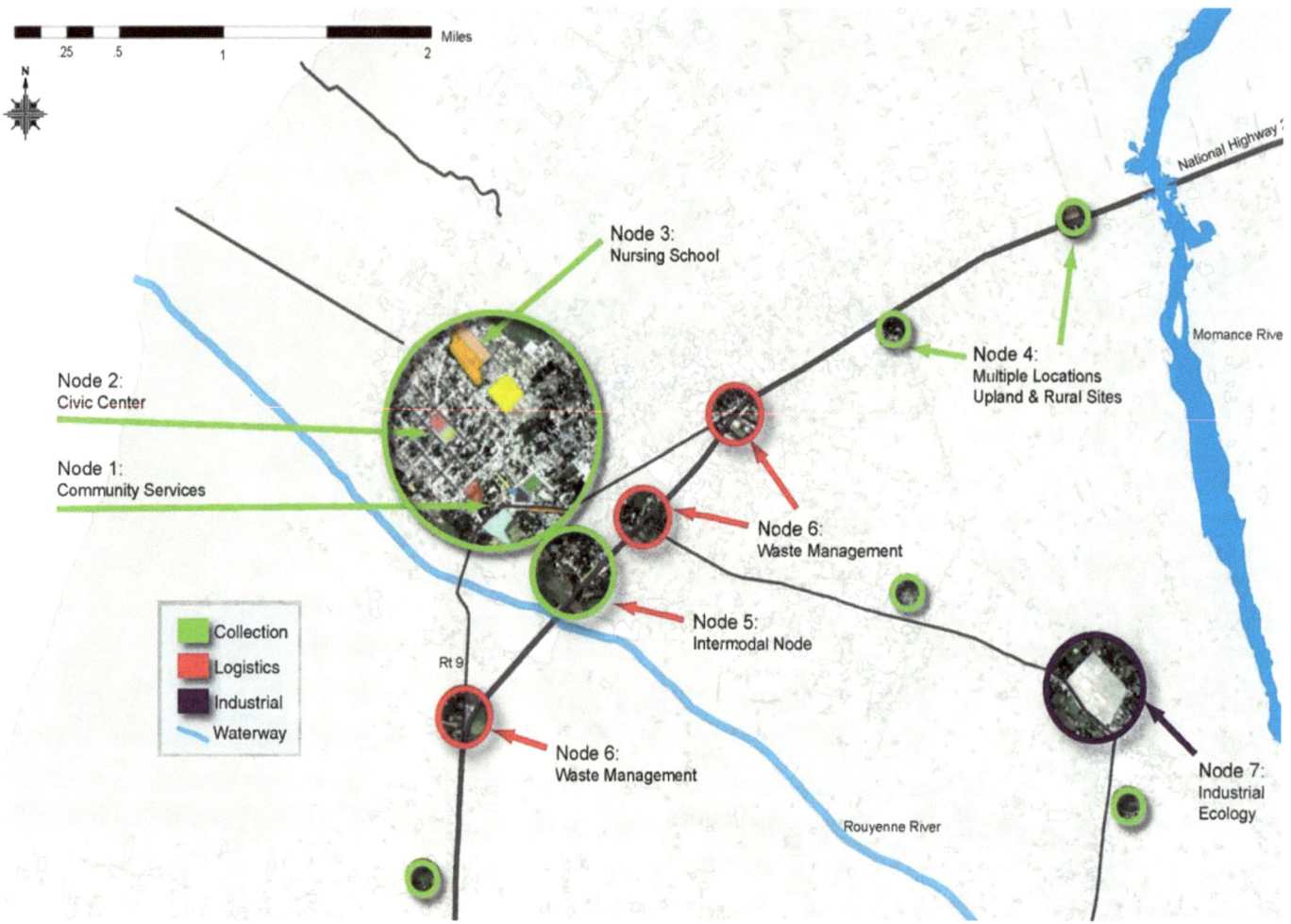

Downtown Infrastructural Nodes

problem
Official waste collection systems are lacking in Léogâne and the waste stream is rapidly expanding, polluting waterways, degrading public health and eliminating the potential economic benefits of utilizing this otherwise lost material stream. With 75% of the total waste produced per capita being organic, it must be handled in accordance with best practices for conversion to final end-use.

recommendations
Multiple infrastructural "nodes" have been identified as community service points for water, power, food, waste collection and other utilities. Nodes create an integrated web connecting dispersed links across existing and new infrastructure and developed around existing services and functions. The whole waste collection system includes processing at the source, separation and handling, storage, transformation of solids, transport, disposal and energy generation. .

synergies across sectors
Nodes are designed to maximize synergies, making the best use of limited resources and with the goal of transforming them into useful commodities. Nodes are also the catalysts for deployment of community services in key city center locations. Waste management fosters economic revitalization through community involvement and environmental stewardship.

challenges
Social waste habits are slow to change, so incentivizing the system and an educational campaign will be necessary to inculcate the benefits of resource conservation. To address the waste management issue, related systems road system upgrades necessary for transport waste, as well as collection administration will be vital to success.

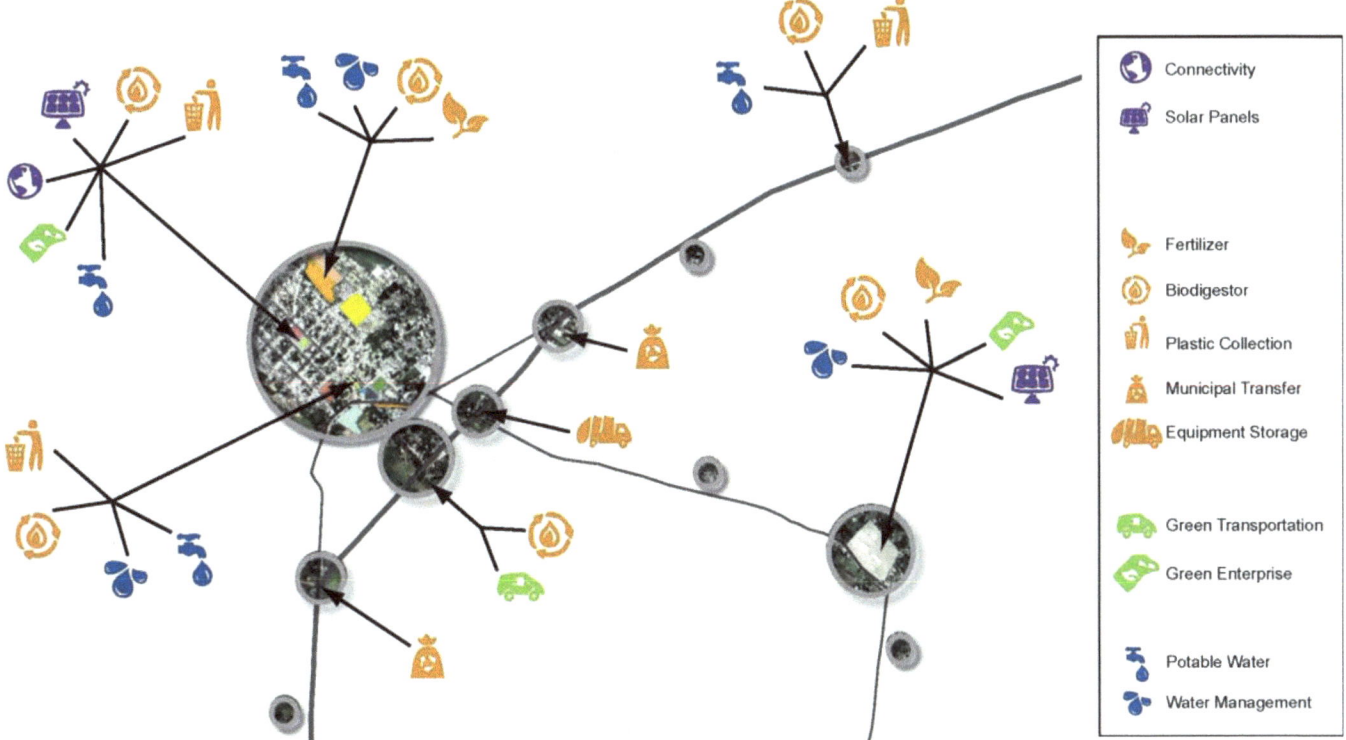

Web Of Infrastructural Nodes

Multiple infrastructural "nodes" have been identified to become the service points for community water, power, food, waste collection and community services. The nodes have been organized according to create a dispersed infrastructure for overall town resource management. These service points function as localized interventions forming an integrated web to connect disperse links across existing and new infrastructure.

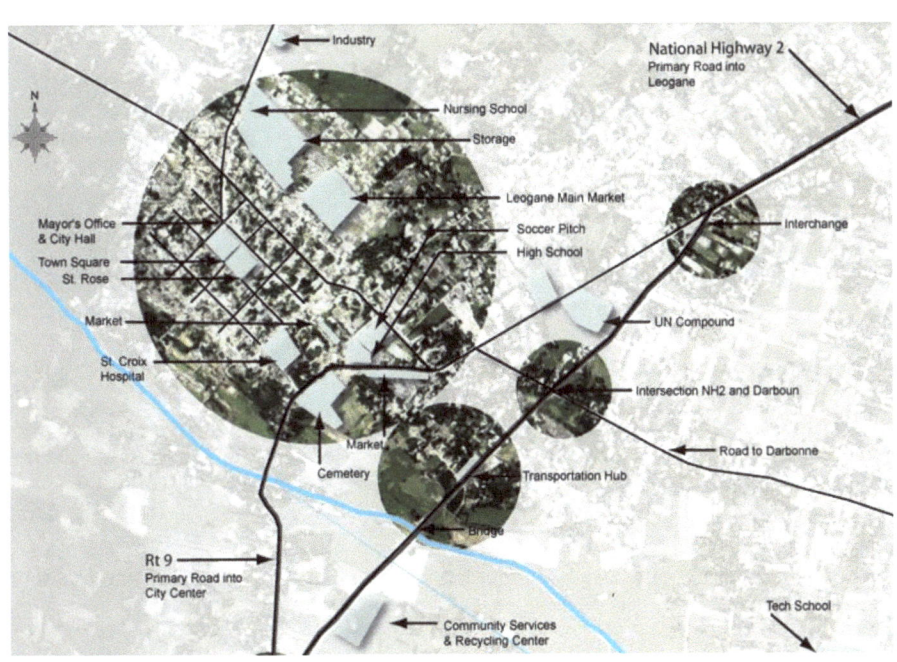

Existing Site Functions

node 1: community hub

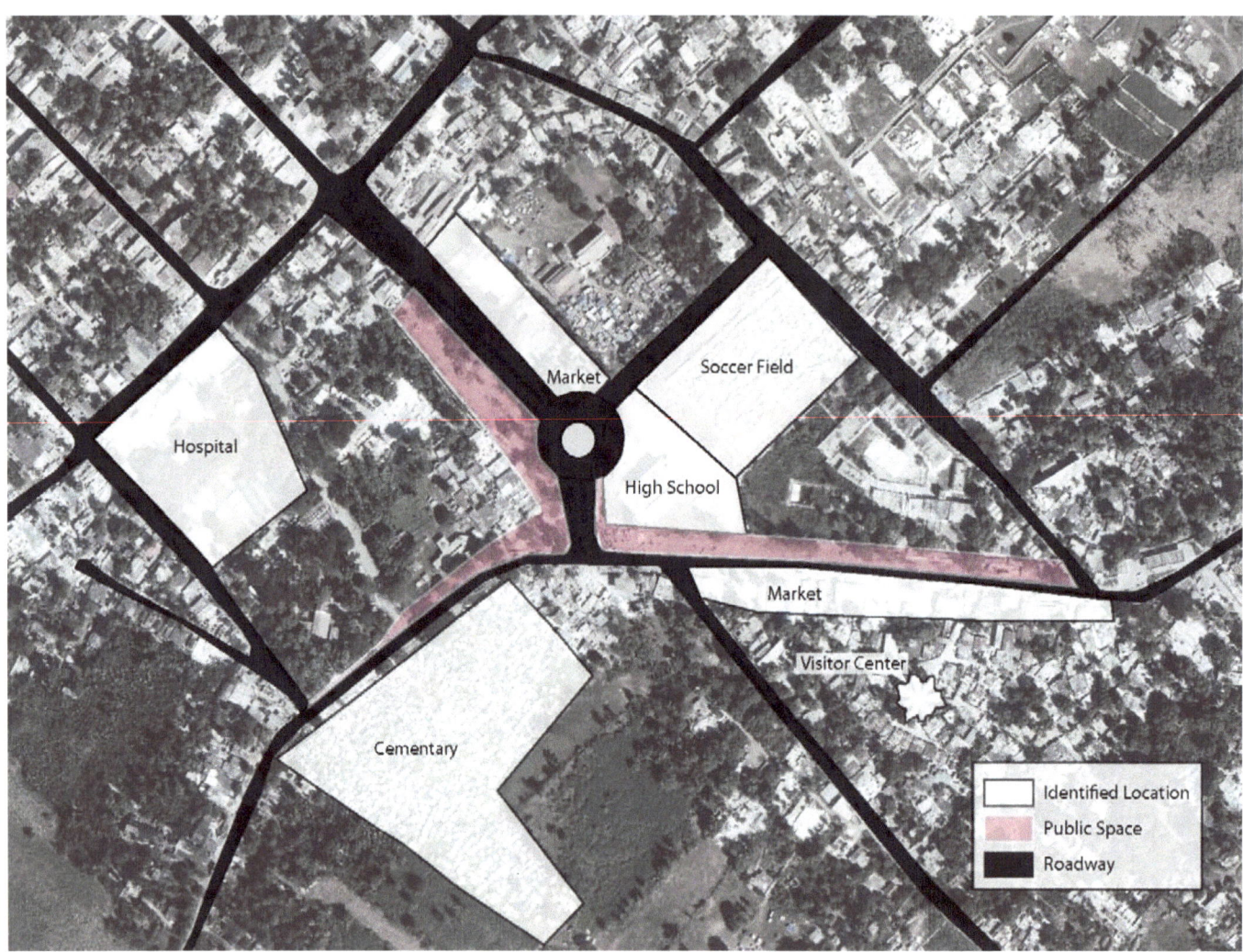

Existing Site Context

problem
The community hub is a functional civic and commercial center without a source of power which limits its economic potential. This locus of activity must be outfitted with amenities to make the best use of the space.

recommendations
In the community hub, biodigestors capable of producing methane from human and organic waste are integrated into the on-site organic waste management plans. The soccer field has been retrofitted with underground shipping containers utilized for water impoundment in rain events. These elements become part of an system for utilizing on-site waste integrated seamlessly through community services.

synergies across sectors
The natural gas from the biodigestor is diverted to a community kitchen which in turn creates organic waste. The rain catchment from the soccer field (an otherwise well-utilized recreational facility) provides both drinking water (with proper filtration) as well as storm water management. Integrating community services with underutilized resources fosters stewardship while building resiliency.

challenges
While technical challenges are manageable, this node demands upfront investment along with technicians to service and maintain the biodigestor and rainwater harvesting facilities.

Shipping Containers used to create an underground space for water impoundment beneath soccer field
Interconnected waste management of sanitary waste and black water
Rain water directed toward rear-wall swales which filter into soccer field storage

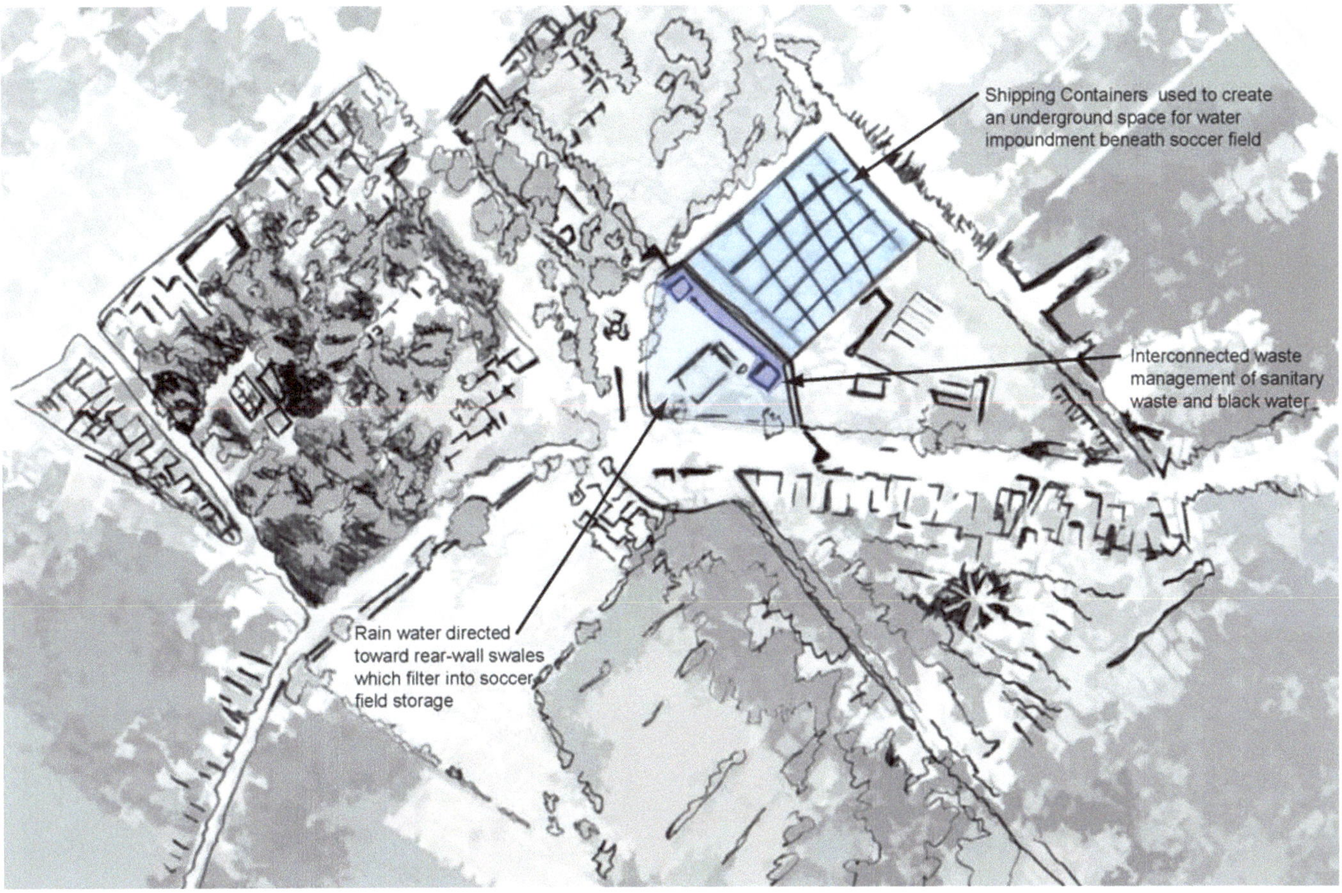

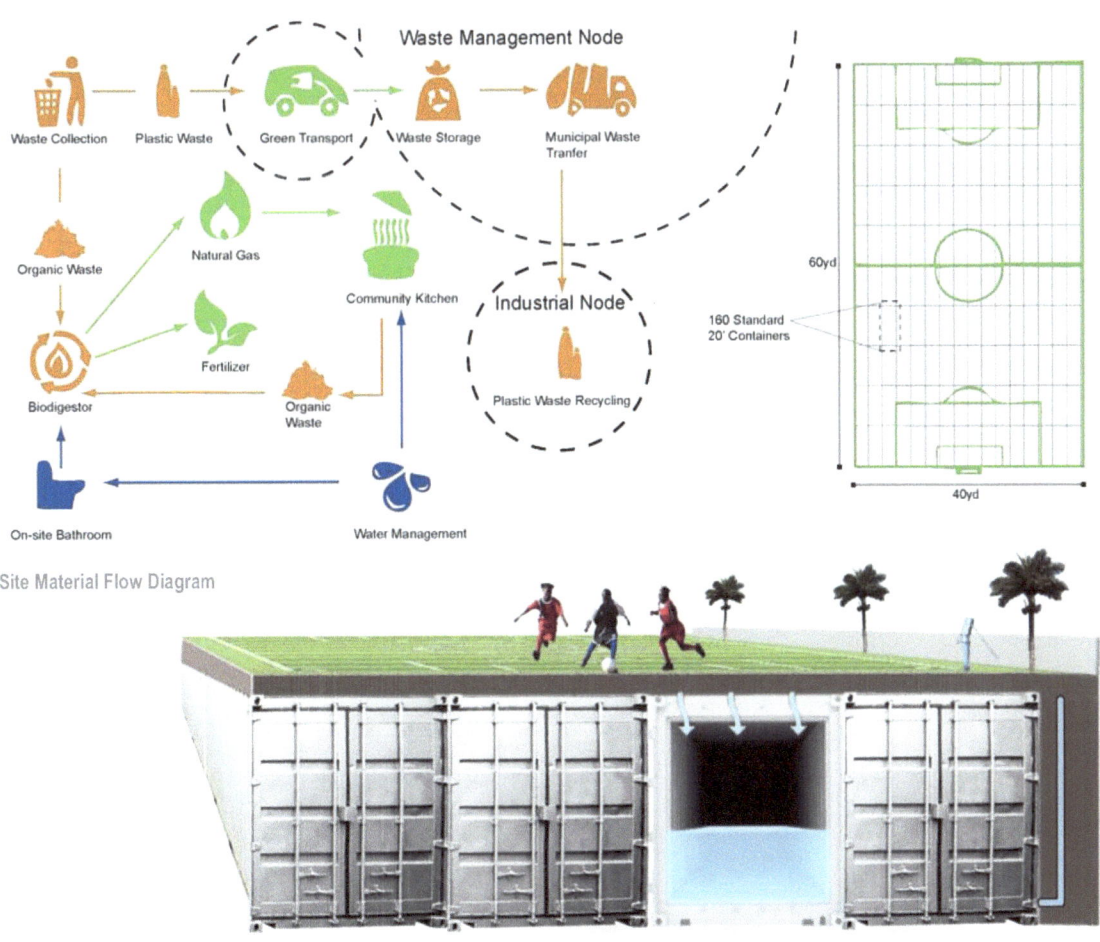

Community Hub Proposed Site Plan

Waste Management Node
Waste Collection
Plastic Waste
Green Transport
Waste Storage
Municipal Waste Tranfer
Organic Waste
Natural Gas
Community Kitchen
Industrial Node
Biodigestor
Fertilizer
Organic Waste
Plastic Waste Recycling
On-site Bathroom
Water Management
60yd
160 Standard 20' Containers
40yd

Site Material Flow Diagram

Stormwater Storage Under Soccer Field in Cisterns for Irrigation

node 2: town square

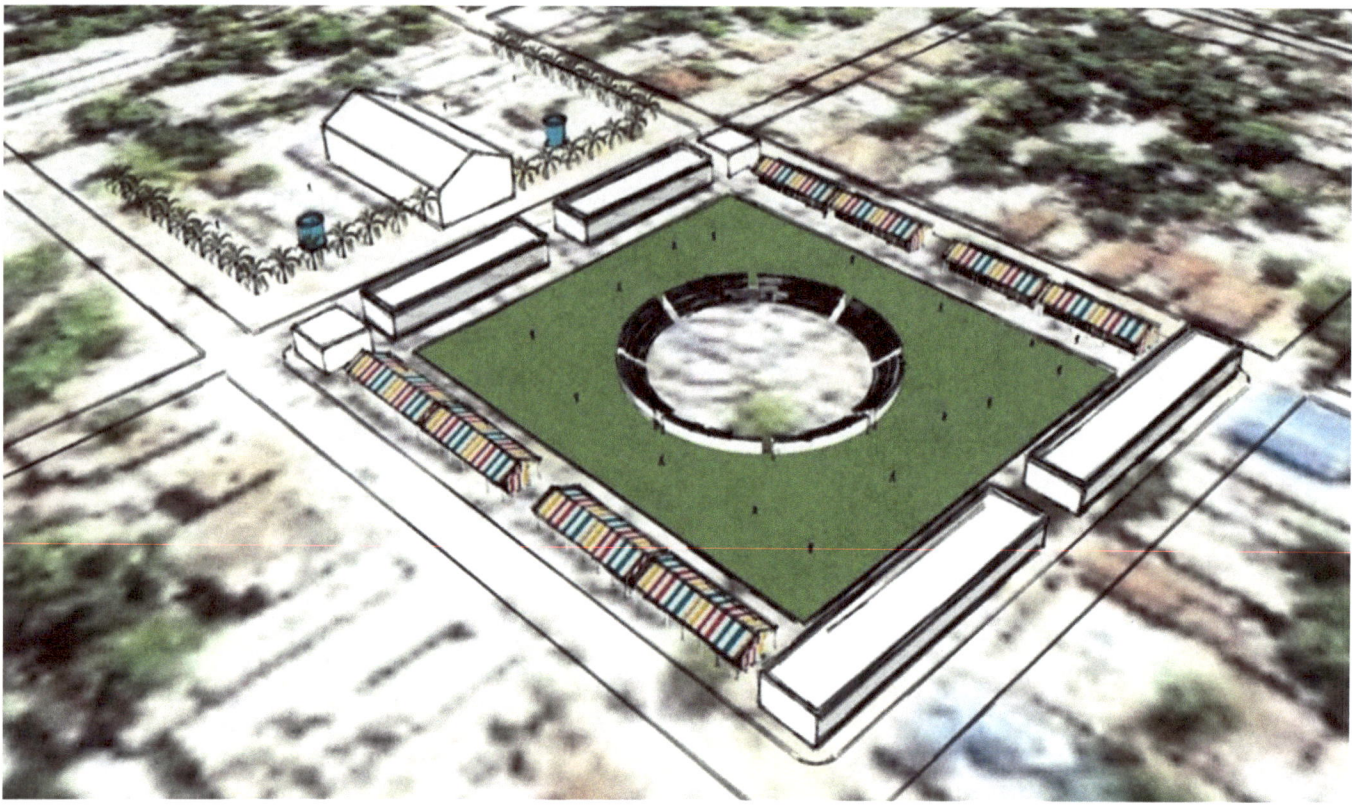

Perspective of Proposed Site

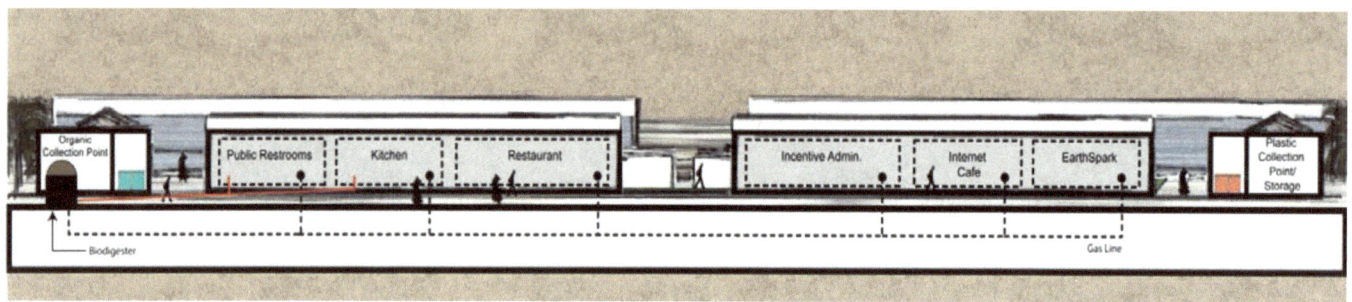

Section with Program Relationships

problem
Civic and commercial functions of this central site, disrupted by tent camps to accommodate displaced individuals following the earthquake, must be rebuilt with new structures.

recommendations
Refocusing additional civic infrastructure around remaining public buildings will create a new market and community center, as well as a centralized waste collection point for the local area. Plastic waste is collected on site and transported to Darbonne, with organic waste collected to feed the on-site biodigestor. An incentive system is developed to make this vision possible.

synergies across sectors
The biodigestor will generate biogas from the organic waste delivered to the site. This can be used to support a restaurant. The town square will host the 'incentives' administration office, proposed community classrooms, catering to art and waste education, an internet cafe, public restrooms and 'sustainable cooking-fuel storefront. Organic waste generated in the market will feed the biodigestor

challenges
Locals must be properly incentivized to bringing waste to this central location, requiring not only 'hard-ware to dispense payouts but education on collection standards for the system operate effectively.

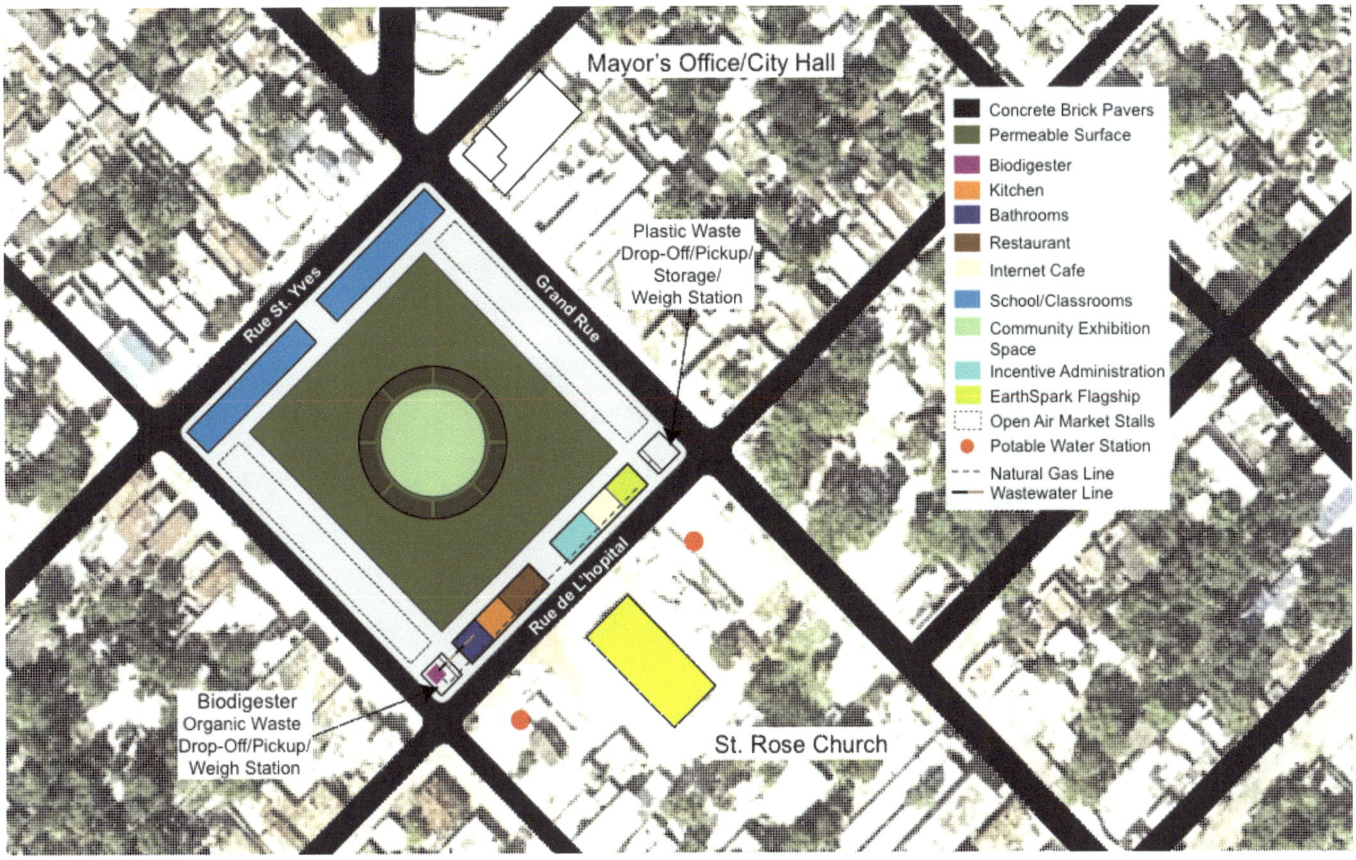

New Market Center and Civic Buildings Plan

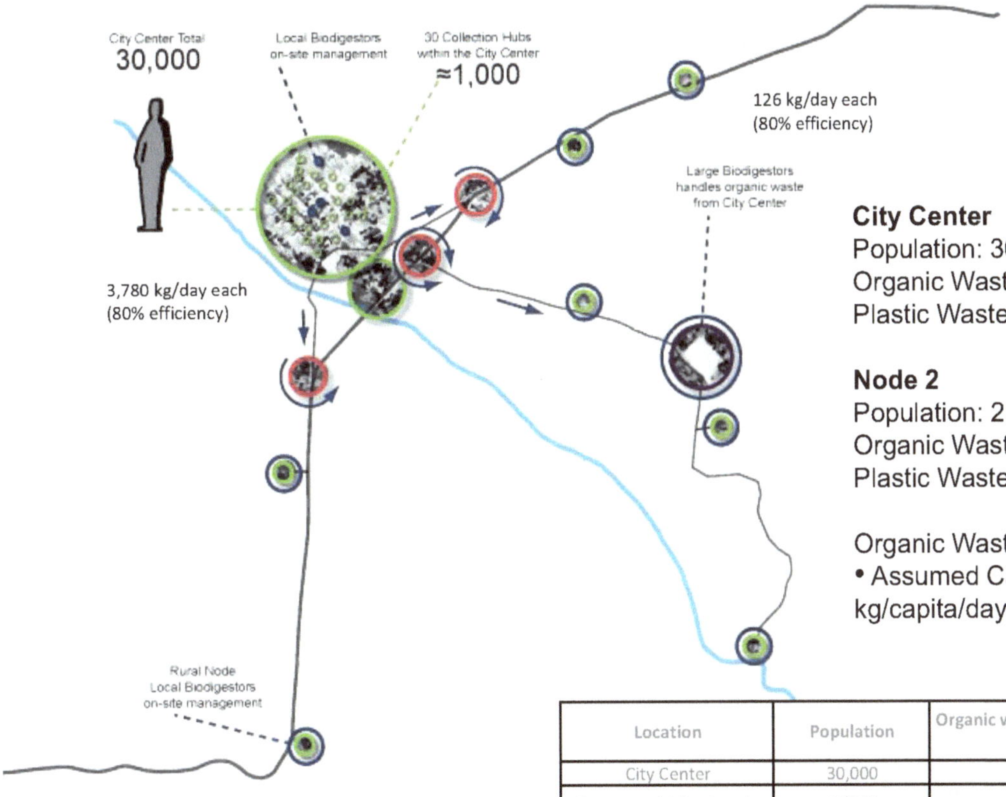

City Center
Population: 30,000
Organic Waste: 80%; 3,780 kg/day
Plastic Waste: 350.55 kg/day

Node 2
Population: 2,000 people
Organic Waste: 252 kg/day
Plastic Waste: 23.37 kg/day

Organic Waste Generation
• Assumed Capture Rate: 80%; 12,600 kg/capita/day

Location	Population	Organic waste Generated Kg/Day	Collection with 80% Effecency kg/day
City Center	30,000	4725	3780
Prototype Upland node	10,000	1575	1260
Upland node	60,000	9450	7560
SUM	100,000	15750	12600

Sources: Philippe F. and Culot M., 2009.

Organic Waste Logistics and Quantification

node 3: nursing school campus

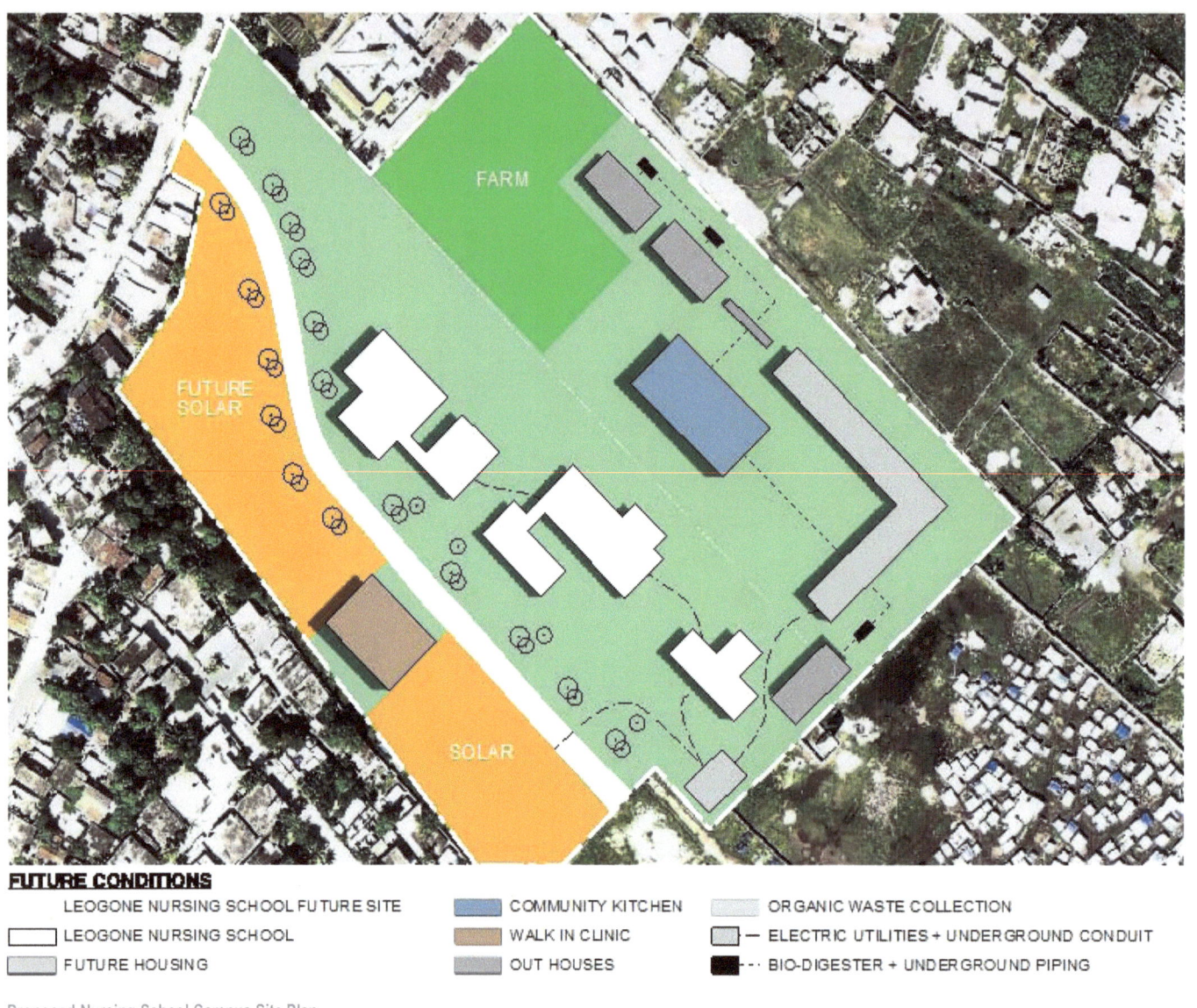

FUTURE CONDITIONS
- LEOGONE NURSING SCHOOL FUTURE SITE
- LEOGONE NURSING SCHOOL
- FUTURE HOUSING
- COMMUNITY KITCHEN
- WALK IN CLINIC
- OUT HOUSES
- ORGANIC WASTE COLLECTION
- ELECTRIC UTILITIES + UNDERGROUND CONDUIT
- BIO-DIGESTER + UNDERGROUND PIPING

Proposed Nursing School Campus Site Plan

problem
Nursing students travel a great distance to this site. It is not uncommon for students to eat only one meal in a day and have limited opportunity to rest and refresh.

recommendations
An integrated campus, with dormitories, a cafeteria and other amenities will provide services and a campus atmosphere to support the students and staff. Organic waste collection systems will collect food waste for conversion to cooking power and gas. Flush toilets will directly feed the biodigestors; grey water will be collected from rooftops and solar arrays provide energy.

synergies across sectors
This low-impact, closed loop system takes natural, organic resources for conversion to site cooking fuel and energy.

challenges
Significant changes in on-site waste management practices and technical staff are necessary for this cooperative system to function. Upfront technological investments and on-going administration will be needed.

node 4: upland rural node

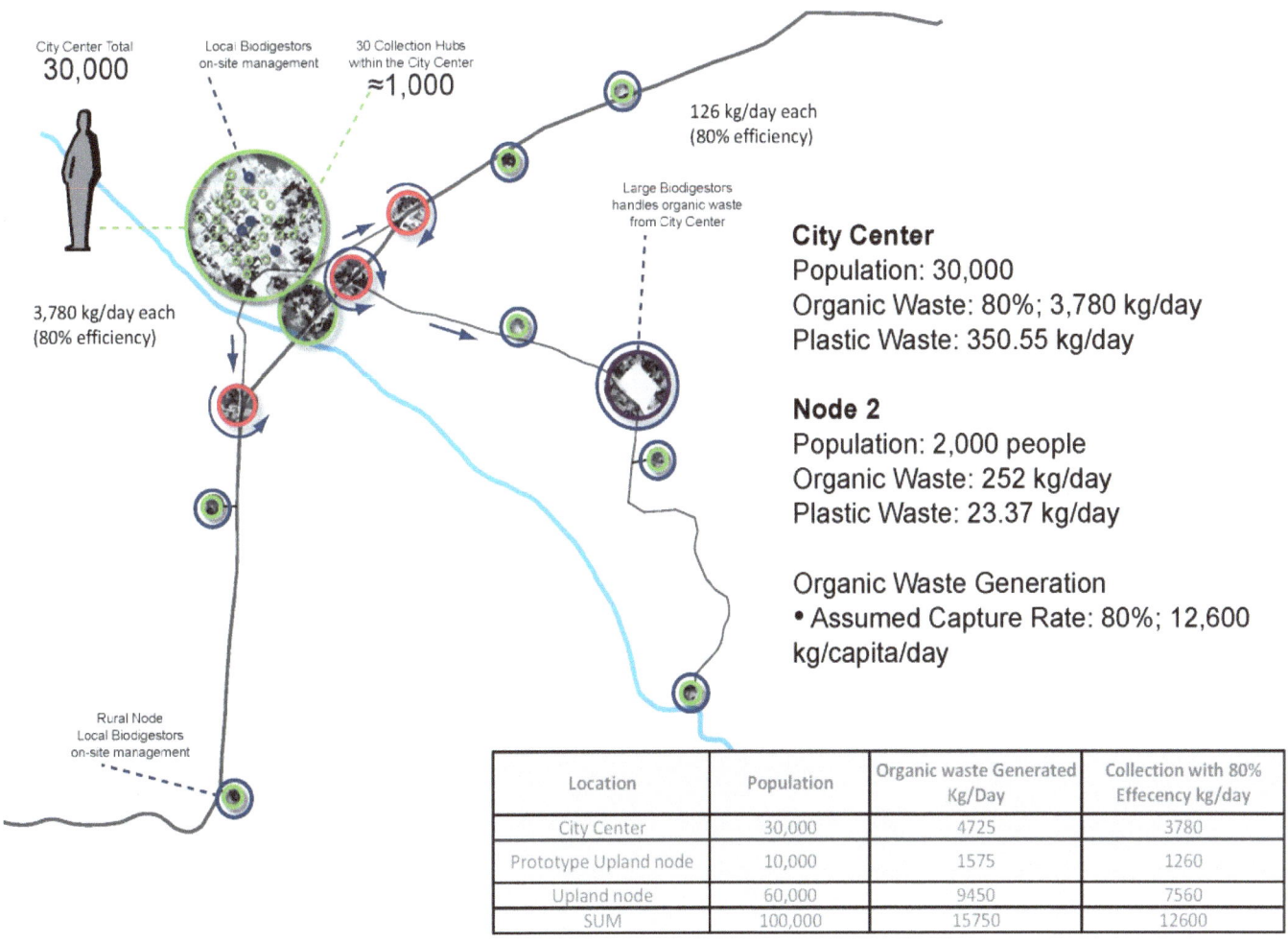

Plastic Waste Logistics and Quantification

problem
Abundant, uncollected waste is periodically set on fire by the locals in order to manage waste. Plastics along with other household waste, such as human and animal excrement, clog canals and prevent proper drainage, creating major health hazards in the surrounding areas.

recommendations
Nodes have a water station, charging station, internet café and public restrooms which can be used in exchange for plastic waste. Such facilities should be located in areas with high traffic that are easily accessible by the dispersed population. Plastics would be accepted in upland locations and then transferred to the main facility in Darbonne to be transformed into bricks. Plastics accepted in upland locations are transferred to a main facility in Darbonne and transformed into 'bricks'.

synergies across sectors
Plastic recycling demands manual labor for separation, an abundant resource in Haiti. In the Darbonne 'eco-industrial park, plastics are formed into bricks by a simple compressing process (no thermal or chemical composition changes necessary). Bricks are sold for the construction of compound walls, retaining walls and barns and otherwise assist in the rebuilding process while incentivizing clean up and sanitation. These collection nodes will provide jobs, supply resources to new industry while delivering basic critical services in remote areas.

challenges
Systems of incentives are need to foster cleanup and change perception of the value of plastics and other waste resources. Payment for waste collected per pound provides a needed incentive, but the system needs a investor to build a market for the new plastic product.

The Upland Node has been designed around an existing infrastucture to minimize construction time and cost

Upland Node Context

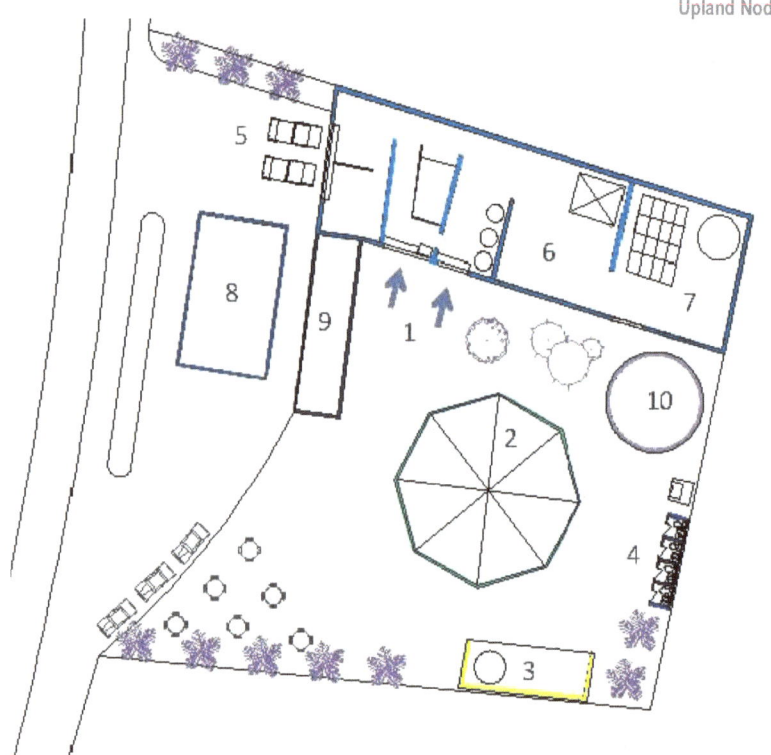

1- Plastic and Organic Collection
2- Internet Café / Charging Station
3- Water Station
4- Public Toilets
5- MRFT Truck Pick-up / Drop-off
6- Plastic Compressing
7- Storage
8- Existing Gas Station
9- Existing Store
10- Biodigester

◦ Rooftop area: 100' x 30'
◦ Total number of panels: 75 panels (3' x 5')
◦ Per hour generation per panel: 230 watts
◦ System's max capacity: 12 kW

Upland Node Plan

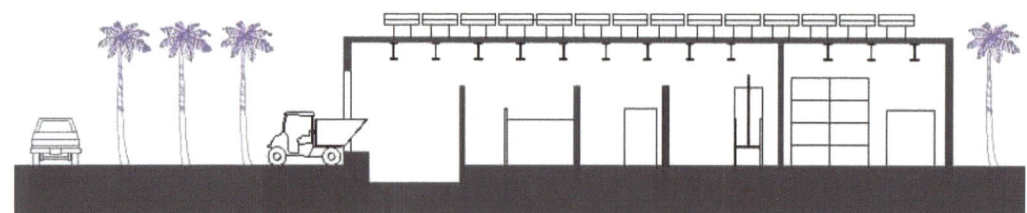

Upland Node Section

node 5: intermodal hub

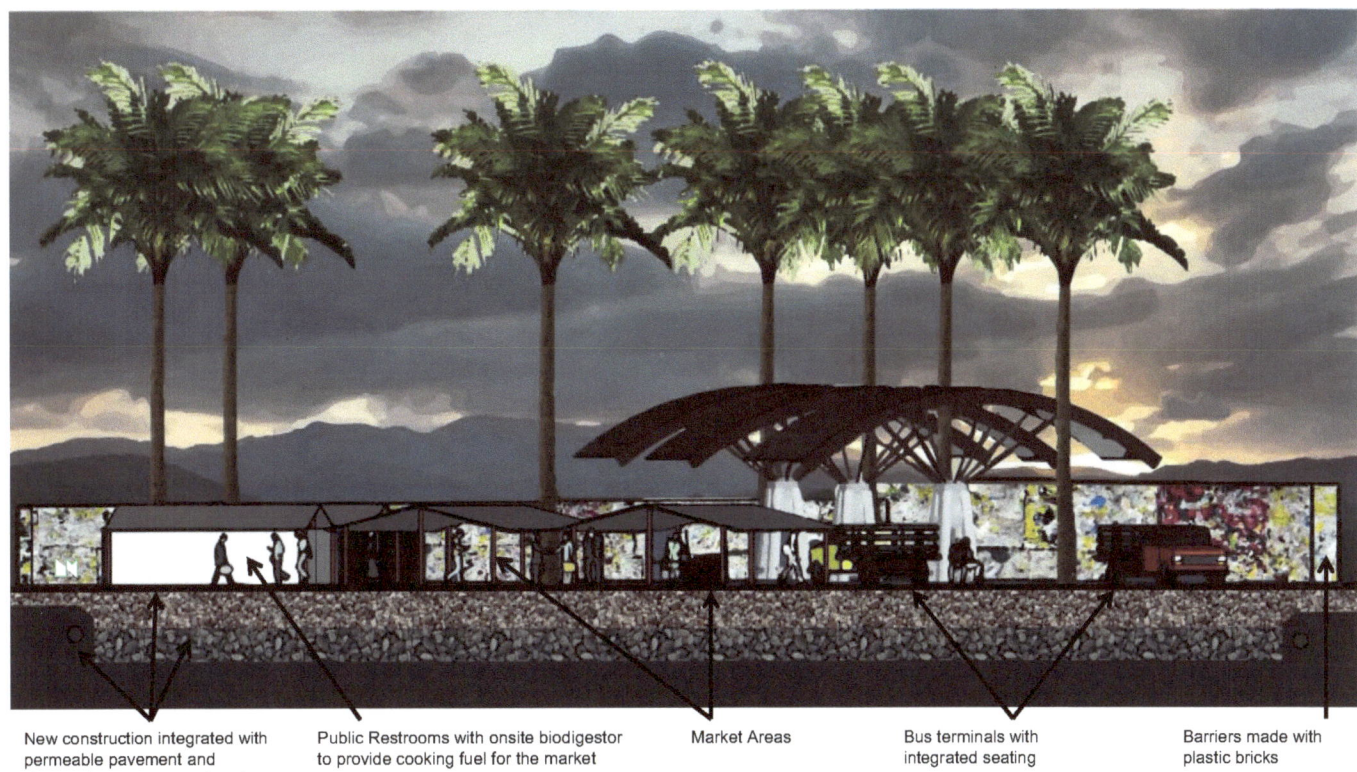

New construction integrated with permeable pavement and stormwater management system

Public Restrooms with onsite biodigestor to provide cooking fuel for the market

Market Areas

Bus terminals with integrated seating

Barriers made with plastic bricks

Elevation of New Intermodal Center

problem
The existing bus stop does not have an organized entrance or exit and lacks waiting space. It fosters a great deal of congestion and limits the utility and effectiveness of public transportation.

recommendations
Development of a multi-modal transportation hub with prioritized access for pedestrians coming from the city center will greatly improve transit function in Léogâne. The hub integrates access to buses, motorcycles, personal motor vehicles, bicycles and pedestrians while providing safety improvements. On-site waste is managed through biodigestors and water through a local storm water management system.

synergies across sectors
Multiple functions are overlaid in the transit hub. The back area of the terminal is designed with a designated pick-up and drop-off areas, market space, waiting area and public restrooms with an attached biodigestor. The biodigestor will be used to fuel the cooking stations in the market area and depending on the amount of traffic in the station, it may have the potential be used as fuel for buses in the future.

challenges
Road conditions are severely deteriorated and lack designated areas for pedestrians. A regulatory body is necessary for public safety to be instilled, driver training standardized and buses properly maintained.

node 6: road improvements and transfer station

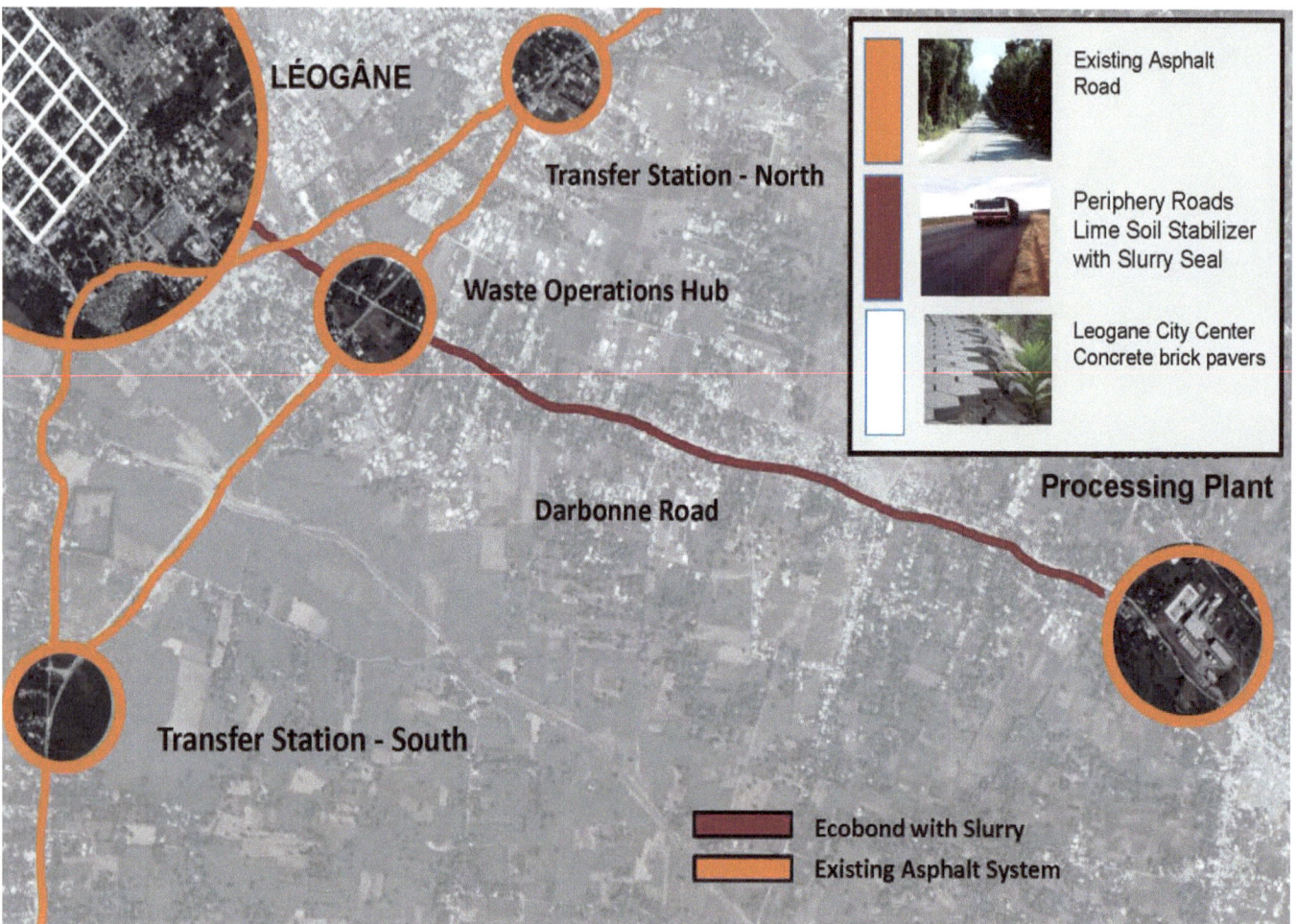

Waste Collection Network

problem
Léogâne's existing inner-city and immediately outlying road conditions are degraded. No sidewalks exist and roads are rutted, with water-pooling supporting water born disease vectors and mosquito nesting. Without a shoulder, accidents are frequent

recommendations
For the city center, hexagonal concrete pavers along main and side streets are recommended as water permeable surfaces easy to repair. On the peripheral roads, line-based soil stabilizers will provide soil stabilization with low environmental impact. With roads stabilized, waste collection in the city center will rely on small vehicles to deposit collections in a municipal transfer facility for transport to a processing station.

synergies across sectors
The waste management will support sanitation efforts in the city and the countryside while creating a potential resource. Roadbed improvement with stormwater drainage helps in erosion control and facilitates economic development with safe thoroughfares.

challenges
Significant changes in transportation infrastructure will be required to aid waste management practices that can help redefine waste as a resource. Centralized management and enforcement of laws that define transportation are imperative.

Separation of waste on site: Organic/Plastic

Residents bring their trash to proposed Organic/Plastics collection points

Residents compensated with **Community Services/Water/Sim Cards** or **Mobile Minutes** in exchange for waste at collection weigh station per kilogram

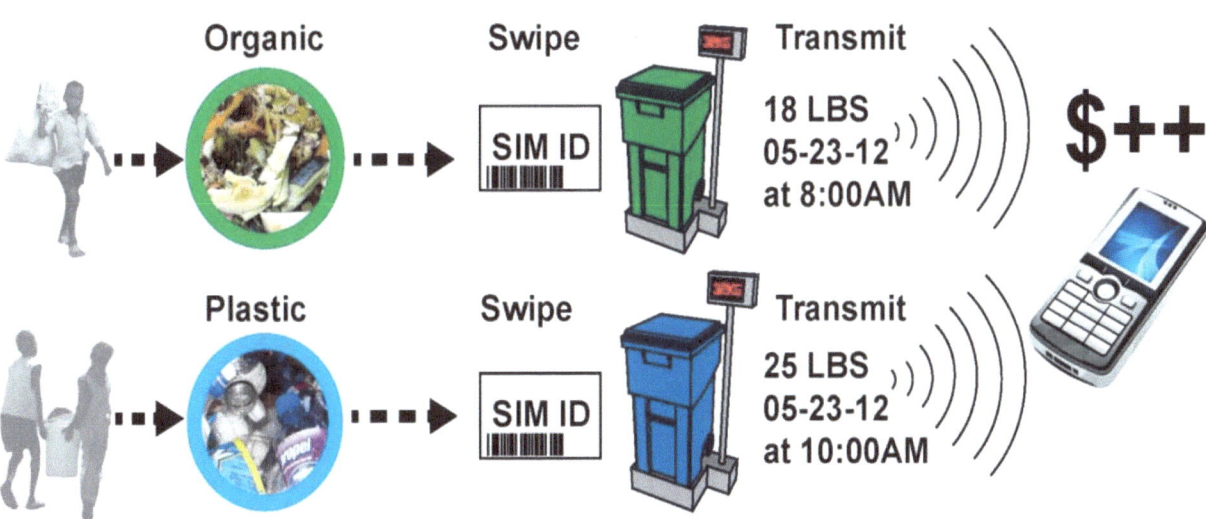

Waste Collection Incentivization

Waste Collection Points

Each Collection point has 4 96 gallon containers.

2 Plastic
2 Organic

They weigh refuse and transmit to central receiving office, time date and identity bar code of SIM card holder. After the waste is transferred the SIM card benefit is rewarded monthly

Light vehicles collect from these bins and move to transfer stations for transport

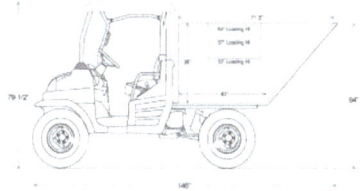

Kubota 3 Cylinder, 21.6 HP Diesel Engine

4WD Operation, 25 MPH

120 MPG Fuel efficiency

Empty weight 2,400 pounds

Hopper capacity 2.5 cubic yards,

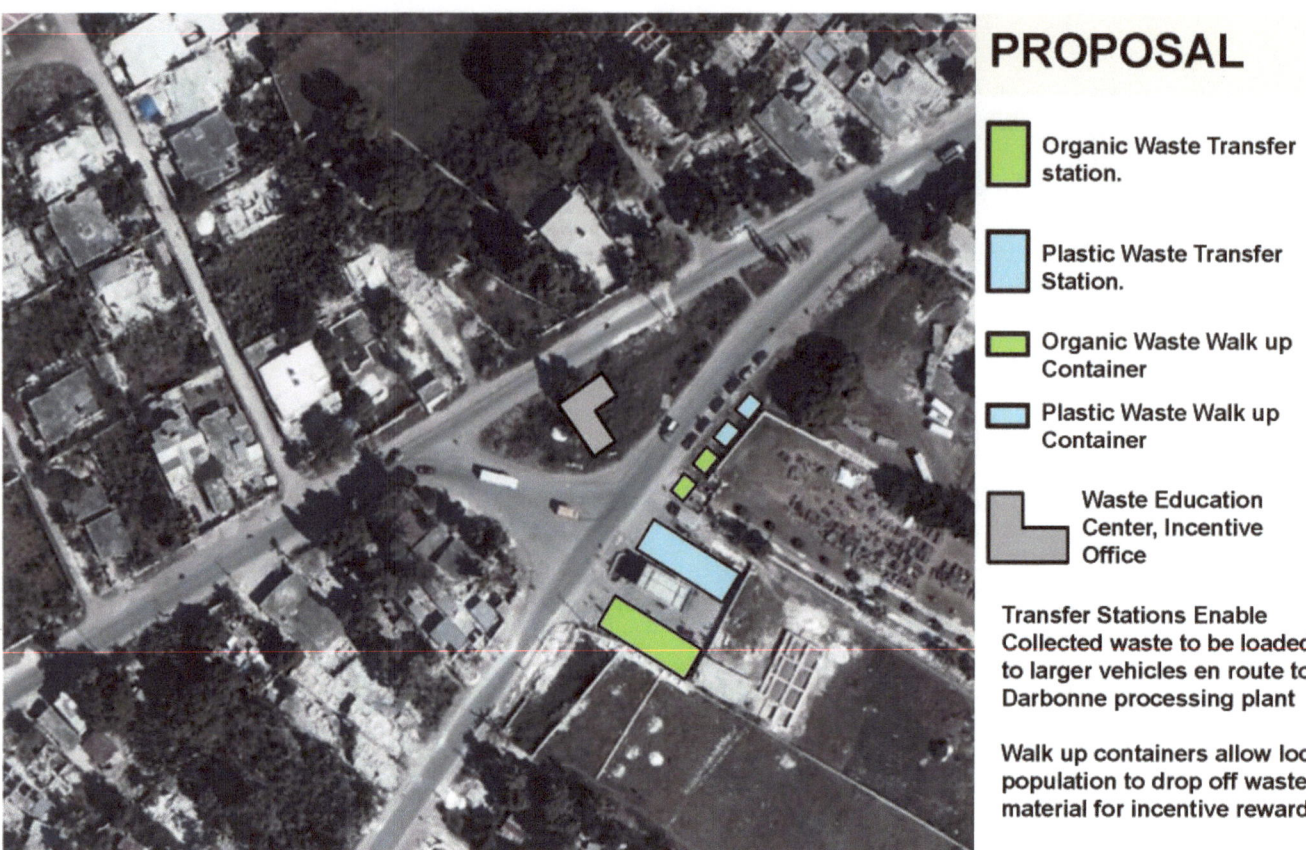

Transfer Station North

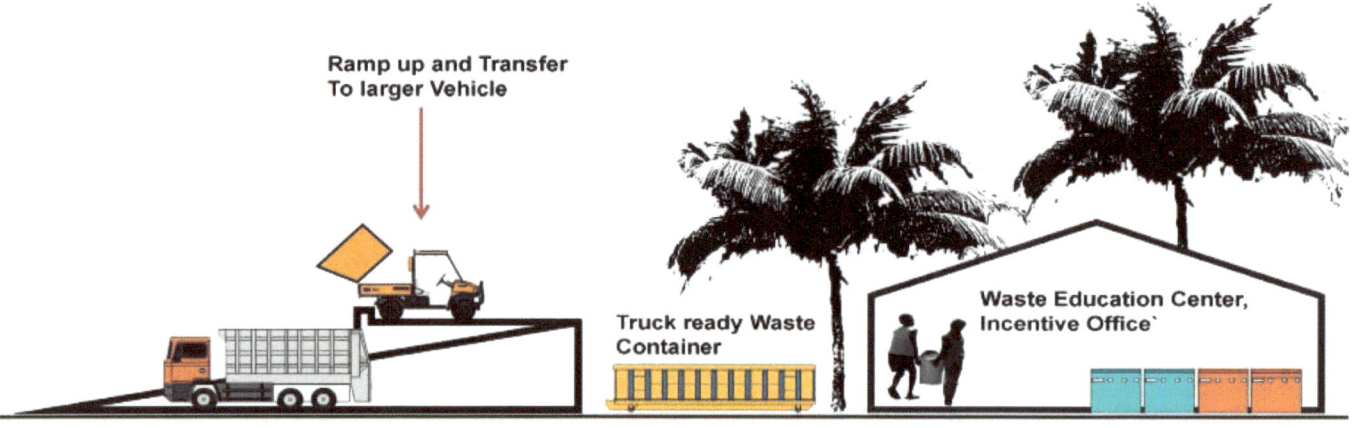

Section through Transfer Station

	City waste management	Total Node waste management	Darbonne Sugar Mill	Total for Leogane
Biogas yield (m3/yr)	44,074	202,350	148.3 million	**148.5 million**
Energy potential (GWh/yr)	0.364	1.21	905.3	**907**
Electric output (Gwh/yr)	0.127	0..43	271.9	**272.5**
Plant Equivalent	14.55 KW	48.5 KW	31.04 MW	**31.1 MW**
Heat Production (cogeneration, GWh/yr)	0.200	0.67	498	**499**

Biodigestor Calculations

node 7: eco-industrial park

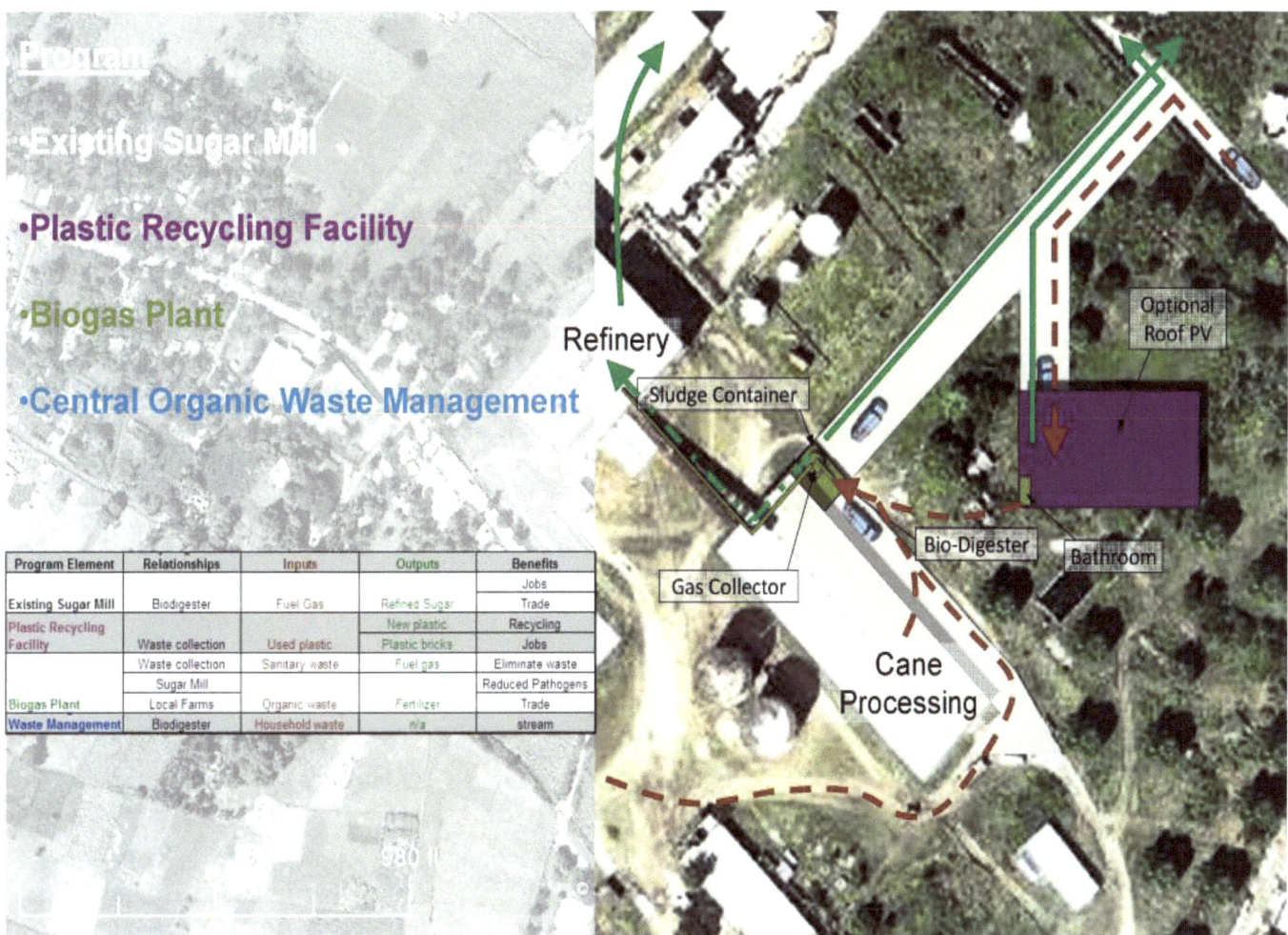

Potential 'Eco-Industrial' park at Existing Darbonne Sugarmil

problem
The Darbonne Sugar Mill is working below capacity with a very low yield rate even though Darbonne is respected throughout Haiti for its sugar production. In Haiti, 75% of municipal solid waste is organic. Significant amounts of organic waste generated per capita are completely under-utilized everywhere as in the city of Léogâne.

recommendations
Revitalizing the sugar cane sector is an important step for Haiti's recovery and can provide much needed jobs, food and energy. Using bagasse, a by-product of the sugar cane crop, can provide secondary or stand-by cogeneration in Léogâne. Bagasse energy must be extracted using a biogas technology, which compared to standard practice reduces pollution, produces high quality fertilizer and has a higher energy capacity. Estimates suggest that if organic waste from the town were diverted for use at Darbonne, the total yearly electric output would be 272 GWh.

synergies across sectors
The upgrades to this technology may be coupled with the facility use for city organic waste disposal. From a series of decentralized sanitation solutions, useful biogas is produced along with agricultural waste and bagasse. Used for cogeneration, biogas manufacture also results in the coproduction of fertilizer. Management of plastic waste on the same site would further centralize waste administration for Léogâne.

challenges
Basic services, such as roads and sanitation must be maintained and financed to support the conversion of the Darbonne area as a major waste management hub. Market analysis is needed to determine overall viability of the integrated system and to identify markets. This requires public engagement and education to alter waste management habits that will beneficially redirect waste flows.

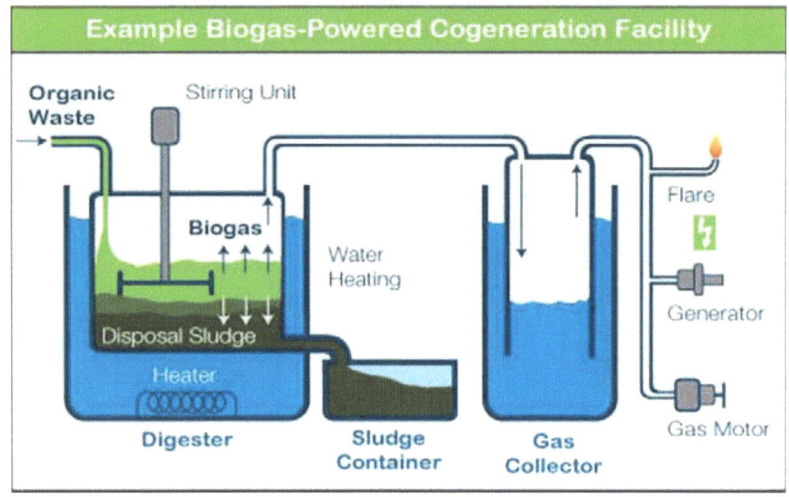

Industrial Scale Biodigestor

Potential locations: sugar mill, dairy farm, centralized organic waste management

◦ Gas for heating or combined heat and power
◦ Requires 'expert construction, operation and maintenance skills

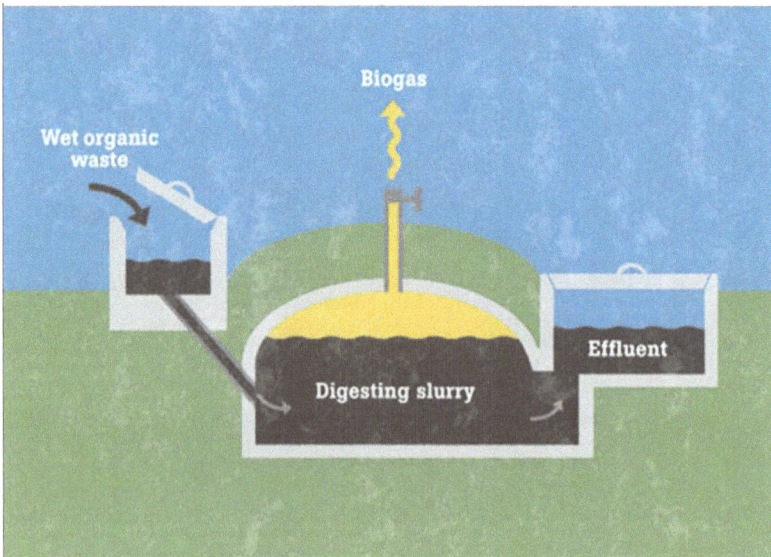

Medium Scale Biodigestor at Darbonne

Potential locations: school, hospital, community center

◦ Dome & plastic digesters
◦ Multiple inlets
◦ Gas for cooking
◦ Green space option

Biodigestor Benefits

◦ Reduce deforestation
◦ Improve sanitation practice
◦ Improve municipal waste management
◦ Smokeless cooking
◦ High quality fertilizer
◦ Reduce GHG emissions

	Burning (optimal conditions)	Biogas
Sugar Production (MT/yr)	100,000	100,000
Bagasse (MT/yr)	215,385	215,385
Biogas yield (million m3/yr)	N/A	148.3
Electric output (Gwh/yr)	133.2	271.9
Plant Equivalent (MW)	**15.2**	**31.04**
Heat Production (cogeneration, GWh/yr)	245	498

Energy Output Burning vs. Biogas

If the organic waste generated in the city is used at Darbonne for producing electricity, the total yearly electric output would be 272 GWh/yr

This is equal to the recorded electric consumption of about 8 million Haitians in 2009 (off-grid generation not taken into account)

Key assumptions based on literature review:

- Ideal sugar production (100,000 MT/yr)
- Bagasse to sugar production ratio (2.15)
- Biogas conversion ratio (0.85m3/kgVS)
- Thermal efficiency (30%)
- Heat production efficiency (55%)

citations and attributions

3	Haiti Overview	Alberto, Dominic; Capobianco, Rocco
4	Integration of Critical Infrastructure ...	Pena, Ana C.; Ward, Miriam
5	Upland Intervention Sites	Pena, Ana C.
6	Upland Site Overview	Owens, Ayan
	Section of Power System	Plaat, Daniel
7	Wind Turbine Site Plan	Plaat, Daniel
8	Wind Site 1	Plaat, Daniel
	Wind Site 2	Plaat, Daniel
	Turbine Elevation and Plan	Plaat, Daniel
9	40 Acre Solar Farm	Pena, Ana C.
10	Floating Solar Array	Todd Weedy, NY Times
	Example of Upper Reservoir	Gu, Jie
	Solar Panels with Agriculture	Korb, Heatther; Owens, Ayan
11	Energy System Co-location Benefits	Pena, Ana C.
12	Rendering of Pump Water System	Gu, Jie
	Pumped Hydro Storage Plan and Elevation	Gu, Jie
	Bird's Eye View of Pump Water System	Gu, Jie
	Horizontal and Vertical Distance ...	Gu, Jie
13	Proposed Reservoirs	Korb, Heatther; Pena, Ana C.; Velivasaki, Artemis
14	Momance River Delta	Pena, Ana C.; Velivasaki, Artemis
	Reservoir Capacity	Korb, Heatther; Pena, Ana C.; Velivasaki, Artemis
15	Shoreline Interventions	Rosas, Maria I. Bueno
16	Zone 1: Arial View	Bliham, Rodger
	"Reef Ball" NGO System	Reef Ball, 2011. http://www.reefball.org/
	Reef Ball Mold System	Reef Ball, 2011. http://www.reefball.org/
	Zone 2: Momance River Delta	NASA; Digital Globe
17	Riparian Strategy: Patch, Corridor, Matrix	Kacker, Priya
18	Riparian Buffer Layout and Illustration	Pena, Ana C.; Kacker, Priya
	Flooding and Temperature Control	Kacker, Priya
	Ecological Services	Riparian Buffer, http://en.wikipedia.org/
19	Application of Bank Stabilization ...	Pena, Ana C.
20	Riparian Buffer Physiology	Haley Heard, *Riparian Urbanism*, Massachusetts Institute of Technology, 2010. http://www.asla.org/2010studentawards/400.html
21	Permaculture Ecovillage Compound	Kinariwala, Danish; Korb, Heather
22	Proposed Permaculture Ecovillage...	Kinariwala, Danish; Korb, Heather
	Permaculture Site Overview	Kinariwala, Danish; Korb, Heather
	Permaculture Ecovillage Compound	Kinariwala, Danish; Korb, Heather
	Compounds Arrayed Around Man-Made...	Kinariwala, Danish; Korb, Heather
	Compound Layout	Kinariwala, Danish; Korb, Heather
	Compound Perspective	Kinariwala, Danish; Korb, Heather
23	Downtown Infrastructural Nodes	Ward, Miriam
24	Web Of Infrastructural Nodes	Ward, Miriam
	Existing Site Functions	Ward, Miriam
25	Existing Site Context	Ward, Miriam
26	Community Hub Proposed Site Plan	Ward, Miriam
	Site Material Flow Diagram	Ward, Miriam
	Stormwater Storage Under Soccer Field...	Ward, Miriam
27	Perspective of Proposed Site	Kahn, Dania; Mauricio, Jessica
	Section with Program Relationships	Kahn, Dania; Mauricio, Jessica
28	New Market Center...Plan	Kahn, Dania; Mauricio, Jessica
	Organic Waste Logistics and Quantification	Kahn, Dania; Ward, Miriam
29	Proposed Nursing School Campus Site ...	Cummings, Steven
30	Plastic Waste Logistics and Quantification	Ivleva, Julia; Ward, Miriam
31	Upland Node Context	Ivleva, Julia
	Upland Node Plan	Ivleva, Julia
	Upland Node Section	Ivleva, Julia
32	Elevation of New Intermodal Center	Morasco, Lisa
33	Waste Collection Network	Celikgil, Nuri
34	Waste Collection Incentivization	Celikgil, Nuri
	Waste Collection Points	Celikgil, Nuri
35	Transfer Station North	Celikgil, Nuri
	Section through Transfer Station	Celikgil, Nuri
	Biodigestor Calculations	Miara, Ariel
36	Potential 'Eco-Industrial' park at Existing ...	Sedita, Chris
37	Industrial Scale Biodigestor	Ameresco Intelligent Systems, 2012. http://www.epsway.com/products-solutions/
	Medium Scale Biodigestor at Darbonne	Ashden, 2012. http://www.ashden.org/biogas
	Energy Output Burning vs. Biogas	Miara, Ariel

© 2012, All rights reserved.

www.ingramcontent.com/pod-product-compliance
Lightning Source LLC
Chambersburg PA
CBHW051109180526
45172CB00002B/836